GIS for Enhanced Electric Utility Performance

Recent Artech House Titles in Power Engineering

The Advanced Smart Grid: Edge Power Driving Sustainability, Andres Carvallo and John Cooper

Battery Management Systems for Large Lithium Ion Battery Packs, Davide Andrea

Battery Power Management for Portable Devices, Yevgen Barsukov and Jinrong Qian

Designing Control Loops for Linear and Switching Power Supplies: A Tutorial Guide, Christophe Basso

Electric Systems Operations: Evolving to the Modern Grid, Mani Vadari

Energy Harvesting for Autonomous Systems, Stephen Beeby and Neil White

GIS for Enhanced Electric Utility Performance, Bill Meehan

Power Line Communications in Practice, Xavier Carcelle

Power System State Estimation, Mukhtar Ahmad

Synergies for Sustainable Energy, Elvin Yüzügüllü

For further information on these and other Artech House titles,including previously considered out-of-print books now available through our In-Print-Forever® (IPF®) program, contact:

Artech House
685 Canton Street
Norwood, MA 02062
Phone: 781-769-9750
Fax: 781-769-6334
e-mail: artech@artechhouse.com

Artech House
16 Sussex Street
London SW1V 4RW UK
Phone: +44 (0)20 7596-8750
Fax: +44 (0)20 7630-0166
e-mail: artech-uk@artechhouse.com

Find us on the World Wide Web at: www.artechhouse.com

GIS for Enhanced Electric Utility Performance

Bill Meehan

ARTECH HOUSE
BOSTON | LONDON
artechhouse.com

Library of Congress Cataloging-in-Publication Data
A catalog record for this book is available from the U.S. Library of Congress.

British Library Cataloguing in Publication Data
A catalog record for this book is available from the British Library.

ISBN-13: 978-1-60807-559-1

Cover design by Vicki Kane

The following figures were provided by Esri, Delorme, NAVTEQ, TomTom, Intermap, increment P corp, GEBCO, USGS, FAO, NPS, NRCAN, GeoBase, Kadaster NL, Ordnance Survey, Esris Japan, METI, and the GIS user community: 1.1, 1.10, 2.2, 2.4, 2.7, 3.6, 3.10, 4.3, 5.3, 5.4, 5.8, 5.10, 5.11, 6.2, 6.3, 6.4, 6.5, 6.7, 6.8, 6.9, 6.11, 7.4, 7.6, 7.7, 7.8, 7.9, 8.4, and 8.8.

10 9 8 7 6 5 4 3 2 1

Contents

Foreword

One of the first projects I worked on, nearly forty years ago, was for a utility. An electric power company hired me to determine the best route, among various alternatives, for a proposed high-voltage transmission line. The challenge was to balance many factors: the cost of construction, impact on the environment, risk of failure of the line, and opposition from various communities. This meant I had to gather hordes of information about the different potential routes. Then, I had to devise a way to assess different factors, such as terrain, access, and proximity to sensitive recreational areas and population centers. I also had to take into consideration the areas that were more susceptible to natural hazards including earthquakes, landslides, lightning, and flooding.

Most of the data I gathered was in the form of maps of every imaginable vintage, shape, scale, accuracy, and authenticity. All of the maps were on paper. Even the unmapped data most often referenced places on the Earth. Everything linked to the geography of the region. The next step of course was to reconcile the various maps and data into a common view, a common map.

Once the data was reconciled, I planned to assign values and weights—from poor to excellent—to each geographic area for each factor. The result of this analysis would be one map that illustrated the best alternative route. At this point, I turned to the computer to help sort out all of the details. I worked with a few colleagues, and we invented a way to organize spatial information in a computer filing system. (There were no database management systems then.) We created algorithms that performed spatial relationships. We devised ways to show the various factors on computer-generated maps. We all discovered that mapping and information technology were meant for each other. Nearly half a century later, the software we developed evolved into what we now call ArcGIS.

Of course, today the world of geographic information systems (GIS) has changed dramatically. Database management systems are commonplace. Smart phones and tablets now display what we once viewed on huge, bulky graphics terminals. Much of the world's computing is performed in the Cloud. As I reflect upon the difficulty of accumulating old maps, the shortage of tools for analysis, and the lack of computing power then, I am amazed that we were able to accomplish all that we did.

Technology has exploded in capability. Content is at everyone's fingertips. And the science of spatial and location analytics is beyond anything we could have imagined even a decade ago. However, the mission of GIS has stayed consistent:

to help solve the world's many problems by showing solutions on a map. This is something no other technology can do.

Maps help us understand our world. GIS provides a way to organize information, perform spatial analysis, visualize patterns that are not obvious, and make connections to solve problems. Maps are perhaps the oldest form of communication known to humankind. Maps are a language unto themselves. Geography matters, whether we are looking for a new place to build a substation or wind farm or we are identifying the riskiest piece of a distribution system.

When I think back to the transmission routing problem, I realize that it was an enormous task back then even with the aid of computers. Today—because we can share information via the Cloud, perform spatial analysis in seconds, and display analysis results on a smart phone or tablet—the solutions to such problems are commonplace.

Bill Meehan's book dives deep into many areas of the electric utility business, drawing connections between the greatest needs of a utility and the many ways GIS can help. Hundreds of workflows are covered in this book, whether you are trying to improve distribution planning, roll out smart grids, or better understand customer behavior. I have known Bill for more than thirty years. He has a unique background with experience running electric operations for a power company, heading up an engineering department, and pioneering the application of GIS to solve utility problems. Bill's passion for GIS and for the electric utility business is contagious and motivational. What I find most interesting about Bill, as evidenced by his many articles, his speeches, and his books, is that he has the ability to make complicated concepts simple.

Finally, his book reminds me of my roots—using geography to help solve a complex utility problem. I recommend this book to anyone who wants to gain insight into the use of GIS technology to enhance the performance of a utility.

Jack Dangermond,
Redlands, CA
May 2013

Preface

A decade ago, I ran the electric operations for a large power company in the northeastern part of the United States. My boss was a guy named Russ. He was the president and the chief operating officer. He was also an accountant. My background was power system engineering. So you would think that our philosophies would be miles apart. Yet, I probably learned more about running a power company from Russ than I ever did from sitting in my Power System Engineering class in grad school. Russ had a few simple rules, and he stuck to them religiously:

- Make money for the shareholders.
- Keep customers happy.
- Keep employees safe.
- Obey all laws and regulations.
- Keep the lights on.
- Respect the environment.

His "goldenest" rule (if there is such a word):

- Keep us out of trouble and out of the newspapers.

If we did all of those things, Russ would argue, we would have done a pretty good job of running a power company. The trick, of course, was to put these guidelines into a practice.

Early in my career, when I was a young intern at the power company, I worked with two engineers who could not be more opposite. Nick was a friendly, helpful old protection engineer. He spent his entire career designing relay protection schemes. As far as I knew, Nick never aspired to do anything else in the world. Apparently, one day, years before I arrived, Nick wrote a three-page memo to his boss detailing his philosophy on relay protection. Back then, memos were typed on old manual typewriters on yellow memo paper. One of the first documents my boss told me to review and memorize was Nick's memo. The subject of the memo was "Protection of the electric system using relays." I discovered that people referred to this document regularly in their daily work. The managers would often ask the substation design engineers if what they were doing complied with Nick's memo. In all the years I worked there, people must have referred to Nick's memo thousands of times. I would have to assume that in his entire forty-plus years of service at the power company, Nick only wrote one memo. (Imagine today only having written one email?) No one ever referred to Nick's memo as his "January 4, 1969, memo," or "Nick's first memo," or "Nick's newest memo," or "Nick's memo about relay protection." No, it was simply "Nick's memo." But obviously it was enough. The performance of the relay protection system was stunning. Rarely was there a misoperation. Nick's memo was simple and profound. When Nick retired, his associate and protégé, Johnny, picked up Nick's work. Johnny followed Nick's philosophy to the letter and as far as I know, never wrote a single memo in this life. He didn't have to. Nick's memo would do just fine.

Dan was the other guy. He was one of the substation design engineers. Dan was arrogant, a know-it-all, and sometimes plain nasty. Dan was a genius of an engineer. Each one of his substation designs was new, unique, and a wonder to behold. The problem was that every substation was different. Each had a different grounding scheme, a different bus-bar arrangement, and a completely different substation control system. He loved to try out the latest substation automation systems, which back then were fairly primitive. The only thing consistent from one substation design to the next was the relay protection schemes. Even Dan wouldn't mess with Nick's schemes. Dan created a maintenance nightmare. When something went wrong, the substation technicians had to scour the wiring diagrams and schematics to figure out what Dan had done. Dan's brilliance and complexities cost the company extra millions, lengthened restoration times, and soured customers.

So what do Dan, Nick, Johnny, and Russ have to do with geographic information systems (GIS) and electric utility performance? Russ challenged me to figure how to deploy technology to solve the big problems that electric utilities face, not to just invent solutions to problems that were challenging and interesting to work on. Engineers like me can get so mired in the details that we forget the relatively simple goals of the utility company. So as I work with utilities to deploy GIS, I try to help companies use this marvelous GIS technology to solve their biggest problems. Often these problems are simple to understand, but hard to solve. What I learned from Nick (and Johnny) was to keep things simple and consistent. As the director of software development at Esri has stated many times to me: simple scales, complex fails. Finally, what I learned from Dan was that great ideas don't often translate into great projects. And, of course, Dan's complex designs tended to fall down over time. Lack of simplicity and consistency can be expensive and disruptive.

My experience over the years working with GIS in utilities, responding to requests for GIS projects, and meeting with thousands of people engaged in GIS is that many of these projects don't really solve the big problems well and are often just too complicated. So, in this book, I will describe ways in which GIS can solve the big problems in simple and consistent ways, because after all what could be simpler than to view business results on a map? People have been doing this for centuries, well before Nick wrote his one and only memo.

What Is This Book About?

GIS for Enhanced Electric Utility Performance describes the how geospatial technology in the form of a modern enterprise geographic information system (GIS) can be applied to all aspects of the electric utility business from smart grid to generation to transmission to distribution to the retail supply of electricity to customers. It is a journey that leads to a full discussion of what spatial technology for electric utilities is all about. This book is for people interested not only in the technical details of a GIS-enabled electric system, but also how such a system works in the real business world. As we examine the complexities of the electric utility business in the book, I will often remind readers of Russ's simple rules, which when you think about it, are the strategic mission of an electric company. I am hopeful that it can explain to us how GIS contributes to the solutions to our biggest problems.

Who Is This Book For?

This book is for information technology managers and executives, GIS managers, system integrators, and application providers. Utility executives, managers, and operating supervisors will also find the book helpful to gain an appreciation of the how GIS is structured and built. This book can also help organizations better manage, integrate, and use their GIS to add real value to their business.

What's Inside?

Historically, a single electric utility handled the entire supply chain of power delivery, production, wholesale delivery, local delivery, and retail. Over the last decade or so, utility regulators have required these vertically integrated utilities to unbundle their business into the component parts. Yet, many of these vertically integrated utilities remain. So the organization of this book is along the lines of a typical vertically integrated utility. Regardless of whether a separate company or a separate division handles generation or transmission, the use of GIS technology and the workflows are the same.

The book has eight chapters, an introduction, and two appendices. The introduction describes what GIS is and provides a glimpse into what the future of GIS may well be, including the use of Cloud computing and the broader use of smart mobile devices. Chapter 1 is an overview of how GIS fits within the utility business model. It gives some history of how electric utilities have used GIS and includes some common barriers to its broader application. The following chapters detail how each major division within a utility has used GIS or could use GIS to enhance their business. They detail energy supply, which includes generation, wholesale or electric transmission, electric distribution, emergency management, customer care, and shared services. Appendix A provides an overview of generic GIS technology, while Appendix B focuses on GIS and smart grid.

How to Use This Book

The book can be read cover to cover to give the reader a complete picture of the breadth of how GIS can transform an electric utility. Or after the Introduction and Chapter 1, readers can skip to certain areas of interest. For example, if utility logistics managers want to understand the role GIS can plan in supply chain, they can jump directly to Chapter 8.

The book contains a list of recommended resources for further study and research. It contains reference books and papers on how real utilities successfully deployed the technology.

Acknowledgments

This book would not be possible without the inspiration of the electric utility customers of GIS around the world who inspired me with their work. It constantly amazes me to see their creativity in the application of spatial technology to their very real problems. The electric utility business is a tough one, with challenges ranging from miserable weather to demanding regulators to economic downturns. Yet given all these challenges, the power gets restored, customers get connected, and power drives the economic engines of the globe. It is the work of the utility employees who rise to these challenges, with GIS playing a growing role.

Special thanks to Chris Kelly from Clearion, Todd Harrington from Union Power, Cindi Salas from Centerpoint Energy, and Trond Ottersland from Powel Corporation for the use of their screenshots. Thanks to Esri and particularly Pete Schreiber for his help in gaining permissions for the Esri screenshots used throughout this book. Thanks to the Smithsonian for permission to use the old Edison map. A special thanks to Linda Hecht and Laura Dangermond for their support. A special thanks to Jack Dangermond, the owner and founder of Esri, for the foreword and for his support on this project. Thanks to Professor Thales M. Papazoglou for his continuous encouragement of my work.

Thanks to the staff at Artech House, and particularly to Deidre Byrne and Samantha Ronan for their patience with me during the acquisitions stage of the book.

Final thanks goes to my loving wife Shelly, who had to endure lonely nights and weekends while I toiled away at this book. I am grateful to her for her patience, understanding, and support, and for those shoulder and neck rubs while I sat at the computer.

Introduction

This bulk of this book deals with business uses of GIS. Given that information technology changes so quickly, a book on the exact technical nature of GIS would likely be out of date as soon as the publisher puts the book to print. For example, more and more software companies are adopting the notion of Software as a Service (SaaS). This means that instead of a company actually installing software on its own servers and computers and upgrading versions of software, SasS to some extent resurrects the old idea of timesharing. In the early days of computing, companies offered users the use of their software and their computers to solve their business needs on a timesharing basis. Users just needed to load their input data (on punched cards) into a card reader that communicated via modem to the timeshare company. Hours or days later, the user got their results on a computer printout. SaaS is the same concept, except the input devices are smart phones, tablets, and PCs using the web. The results are immediate. As is the trend with many platform software provides, GIS is offered as SaaS.

So What Exactly Is a Utility GIS?

Perhaps the best reference guide to GIS is Roger Tomlinson's book *Thinking About GIS*. It answers the question of what exactly a GIS is. While it doesn't deal with utilities, it provides an outstanding and practical guide to the implementation of GIS.

A Computerized or Digital Map?

There are a number of ways to draw a map or illustrate a utility operating map using a computer. Perhaps the simplest way is to scan a paper map into a JPEG file. It's a map and is created by a computer. Anyone can edit, email, and store the file in a database. A user could draw a map from scratch using a simple drawing program that comes free with any PC, tablet, or smart phone. A drafting technician could create a very complex and detailed map with an expensive computer-aided design (CAD) program and store the map drawing in a file. Finally, users can create maps with a GIS.

Each Is Very Different

The scanned image is basically a picture of the old paper map. While it is possible to extract features such as symbols of transformers and poles from the image, the process is not exact or fast, and users could not easily create intelligent features on demand. A simple drawing program does produce unique features such as lines, circles, and text. However, there is no easy way to associate attributes of features with the symbols of the features.

Designers use complex CAD tools to generate detailed design drawings, like structural details, automobile parts, and buildings. People can use CAD tools to create maps for sure. Also they can link attributes to symbols, like lines and circles. So, users can create an inventory of parts from a CAD file.

What differentiates a scanned image, a simple line drawing, and a complex CAD drawing from a GIS is that a GIS is an information system, not a drafting or drawing system. The primary purpose of each of these tools (scanned images, simple drawing programs, and CAD systems) is to automate in some fashion the process of making a drawing using computer technology. Each file they create represents a distinct drawing. A drawing has a boundary, a border, a finite limit to its extent.

When utilities recognized that their sets of maps were getting more and more complicated and required more staff just to maintain them, they looked to computers to help solve the productivity problem of how to create the maps they needed and were used to using with computers. They invented ways to replicate the old maps using computers. If they needed map sheets at 1 in = 400 ft, they created separate files for those sheets. If they needed a map of the same area at 1 in = 50 ft, they created yet another map sheet at that scale. If they needed an overview map at say 1 in =10,000 ft, they created yet another file. Each of these files replicated a map sheet. If they modified one map sheet, they would have to remember to modify all other related map sheets. In effect, they automated a legacy process. Rather than attempting to understand and transform the larger and more strategic business processes that required mapping, they instead attempted to automate the tactical and legacy processes of map making. They retained the concept that maps were simply individual and standalone drawings. Once they automated the legacy mapping process, they simply plotted the maps from the computer, but used the printed maps in the exact same way they used the old paper maps.

Since the GIS is an information system, not a map sheet creation machine, it can create maps on the fly that display new information each time. In effect, the GIS is a spatial data management system, like a database management system (DBMS). Like a DBMS, when someone queries it, it produces the result of the query in the form of a map. In fact, the GIS can produce results of a query in a table as well. It can produce a list of all transformers that the company has not maintained in the last two years in district 1. Unlike a map sheet system, there are no stored drawings. Instead each output is simply the result of a spatial database query. The output can be any scale, using any symbol, without worrying about the borders of a conventional map sheet. It can create a map from the query, "Show me all the power failure locations in the world for 2012 or show me all the power failures in Westerville, Illinois." This query is all within the ability of the GIS—provided, of course, the GIS has the data it needs.

GIS can take information from sources not stored in its data files and include that data as part of the display *as if the data were part of the GIS*. In database terms, this concept is similar to a join. As long as there is some common key, two or more systems can seem as one. Although any feature in the GIS (e.g., a transformer) can be easily integrated with another system if the equipment has a common identifier. Thus, the GIS doesn't have to replicate attributes of a piece of equipment if another system, like a materials system or enterprise asset management (EAM) system, maintains that data. The GIS can access that data, *when and only when it needs it*.

GIS can combine scanned images, rasters and light detection and ranging (LiDAR) imagery as long as the images have the same reference coordinate system as the GIS. As will be discussed a number of times in the text, the notion of combining spatial data sources produce some fascinating results.

It shouldn't matter where the data is stored (in a private cloud, on premise cloud, the open Internet, or in light databases on tablets). Further, it shouldn't matter how users get the data (over the web, on a heavy desktop client, or through web services). The notion that a GIS creates map products just like the old mapping department with its Mylar and linen original maps and India ink pens would be like using a chain saw to cut trees without turning the saw on.

What is a GIS? A casual answer is that it's a computer program that creates maps. People can display those maps on computer screens. Users can plot or print the computer-generated maps. In fact, the maps can look very much like the old-fashioned maps that map makers have created for centuries. The fact that people have relied so heavily on maps suggests that there is something special about a map. It is a universal communication method for figuring out where things or people are located. No one really knows who created the very first map. What we do know is that maps are as popular today as they have ever been. The popularity of consumer mapping apps on smart phones and universal appeal of navigation systems has made most people map and location aware. Figure I.1 illustrates a modern web-based map of an electrical distribution system. Mapping of utility systems is natural.

Certainly a GIS does create maps. At its heart, it is not a map-making machine, a mapping app, or a computer program designed to make maps more easily or to make them clearer. A GIS is an information system whose main purpose is discovery. It manages data organized by location for the purpose of visualization, analysis, and dissemination throughout the organization. It creates new information from raw data. If the primary use of a GIS is to replicate old manual maps, then users are vastly underutilizing it. If instead they use the GIS to transform a business by giving greater insight into a business problem that results in informed decision making, which in turn leads to focused action, then the GIS is doing its job. If they use the GIS to document the location of an underground cable, that's valuable to be sure. If in addition, they use the GIS to discover which cable is most likely to fail, then that's much more valuable.

For example, a GIS could answer this question: "Where are there places in my electric network that a single event could have the most devastating impact?" The GIS results would show a simple map, with a big red circle around the most vulnerable parts of the network. The action resulting from this knowledge is that the utility takes action to mitigate the high risk. So, in this case, the business result

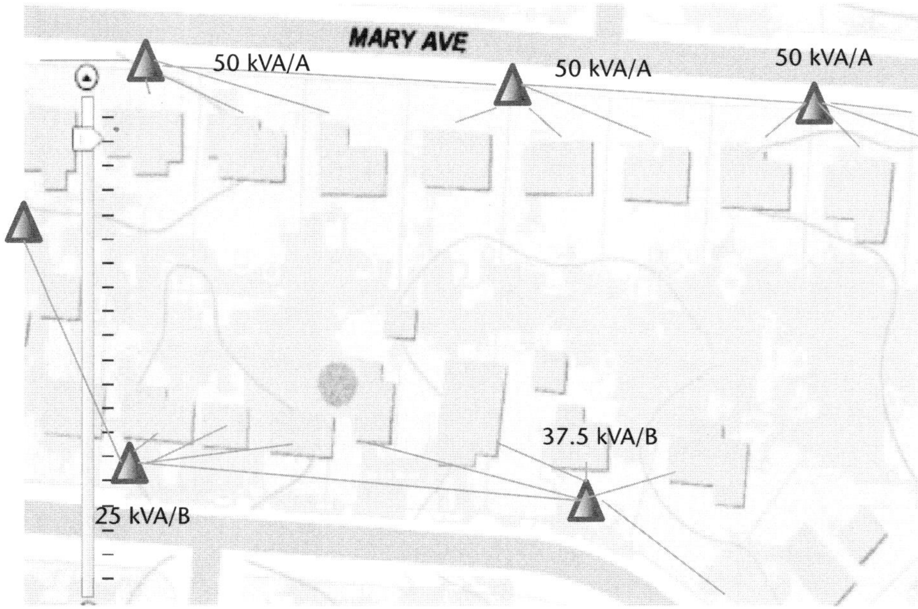

Figure I.1 Typical web-based GIS map showing the electrical distribution system (Source: Esri et al.)

is not to create a better, neater, faster-produced map, but to save the utility lots of money and serve its customers better.

The litmus test of the GIS's impact would be: "Can the utility document improvements in their key performance indicators that are directly attributed to the GIS?" If not, then the GIS has not served its purpose.

Like most information systems, the GIS consists of several parts: a set of software programs (desktop, applications, server software, database management systems, operating systems), hardware (servers, storage systems, the cloud, networks, smart phones, tablets, PCs), and data. Today a GIS is part of the overall information technology (IT) infrastructure that includes the ability of its data and functions to be fully integrated with other key financial, customer, and real-time IT systems at the utility. This is relatively new.

What makes the GIS different from other information systems is the notion of information layers or themes. Figure I.2 illustrates the various layers of information an electric company keeps. These themes or layers can exist in the GIS. Electric companies worry about their electric network, of course, but also about the attitudes of their customers and the sensitive habitat along their transmission right of ways. They may also worry about where they are spending maintenance dollars compared to other areas or where the infrastructure is most vulnerable to terrorist attack or weather events. They worry about where their employees are and where they are making or losing money. The network, supporting structures, customers, land, environmental features, and streets are common layers in a utility GIS. Less common, though interesting, are predictive themes or layers, like flood overlays, climate change impacts, lightning strike history, acid rain contamination, and even bird migration (critical for the site selection of wind turbines). Layers help define

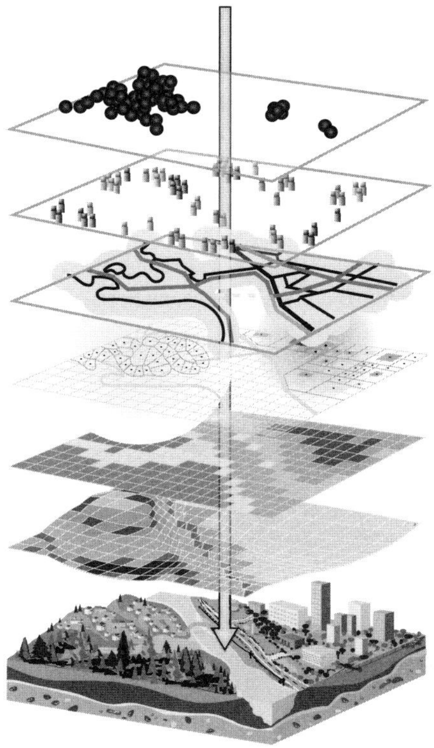

Figure I.2 Typical information layers for an electric utility GIS.

what information products need to be produced to help in the utility's decision making. A layer is analogous to the old system where people made transparent layers for, say, a city for each decade to see how the city progressed through time. Each layer added the changes to the city map.

So just prior to a major hurricane event, for example, the utility might be interested in the layers that illustrate where shelters are located or where flood waters may rise to inundate their critical facilities.

Geocentric Versus Geoenabled

There are two major categories of GIS usage, geocentric and geoenabled. Most utilities built their GIS to capture the essential elements of the electric network—the poles, wires, and transformers. This type of GIS application is geocentric. The work flow involves transactions to the map, particularly the authoring of the components of the map. So any process that creates the features of the map is geocentric. Tools like adding, modifying, and deleting elements in the map are geocentric. In a local government application, adding parcel layers, digitizing easements, and updating ownership attributes of land features are examples of the geocentric usage of the GIS.

Taking Advantage of Mapping

There are dozens of workflows at an electric utility that historically have not leveraged location or the use of maps. For example, a utility may have been performing customer satisfaction surveys for years. For many utilities, customer satisfaction rankings are one of top key performance indicators. The results of those surveys are commonly delivered in a table, spreadsheet, business intelligence (BI) dashboard, or a report. So the customer satisfaction improvement workflow might look like this:

- Hire a company to perform a customer satisfaction survey or participate in a regional customer satisfaction survey.
- Review the results of the survey and compare to a benchmark or to a best-performing utility in the region.
- Develop a plan to improve customer satisfaction.
- Take the survey again and see if the improvement program produced any change.

This workflow misses two important components. First, it doesn't highlight where customer satisfaction is most severe, nor does it identify why customers might be unhappy. Plotting the results of the customer satisfaction in GIS identifies those regions of highest dissatisfaction. Plotting other factors like poor reliability allows the utility to correlate the poor customer satisfaction with other causes. The company could add other factors, like the locations of high bill complaints or unpopular tree trimming activity to gain additional insight. Once the where and the why have been determined, the company can execute a focused program to address the core issues. Figure I.3 illustrates how the GIS can plot the results of a customer satisfaction survey.

This is an example of geoenabling a workflow. Just plotting the results of the survey alone gives some insight, but correlation of many layers of information sharply focuses the analysis and leads utilities to action that is based on science, not just on intuition. Correlation of spatial information is one of the key aspects of a modern GIS. Another example of geoenabling is the analysis of risk of transmission failure. In most cases, utilities determine the risk to the transmissions system by network analysis, like load flows, short circuit, state estimation, and stability. However, these systems do not take into account risk factors such as proximity to small airports, earthquakes, or landslides that could increase the risk of failure and extend the restoration time in the event of a transmission line failure. GIS routinely manages these factors. Adding the results of network analysis with the spatial analytics of the GIS greatly enhances the insight into transmission risk as well as helps focus risk-mitigation activities.

Geocentric Workflows Build Spatial Content

Geocentric workflows have been around for a long time. Most utilities have abandoned the use of paper maps to document their networks and have adopted some form of GIS to do this. Many have also implemented a design process within the GIS. These workflows are geocentric since the GIS directly does much of the work.

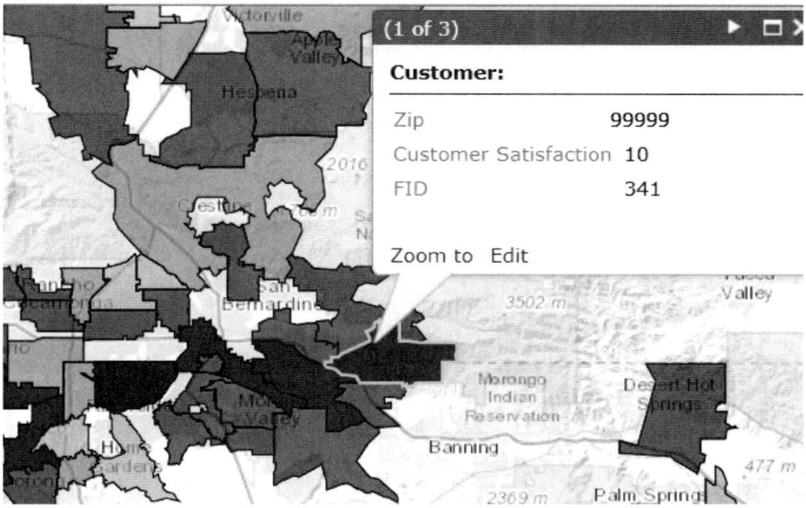

Figure I.3 Example of customer satisfaction shown in the GIS (Source: Esri.)

In these cases, the workflows actually create or author geographic content. Thus, a workflow that creates new or edits geographic content is geocentric.

Geoenabling Enriches Existing Information

A workflow that uses the geographic content to supplement an existing workflow is geoenabled. Just displaying the results of a survey on a map is geoenabled, but to really get the benefit from the GIS, adding spatial analytics to the workflow adds insight and discovery. A common example of advanced analytics is in the analysis of elections. Election analysts plot projected voting patterns as the results pour in, but also add other factors, like demographic data, to get very accurate predictions of what will happen. Figure I.4 gives some examples of geocentric and geoenabled workflows.

The Future of Utility GIS

Client server technologies for GIS will continue for some time now. In this situation, the users operate a heavy client on a desktop computer to add content to the data stored by the data management server of the GIS. Due to the complexities of the data and the detail required to capture all the electric utility assets, users will continue to need the client horsepower to perform these tasks. There will be more evolutionary improvements in the technology. GIS desktop software will become faster, more user-oriented, and easier to use. The modeling of complex networks will continue to improve, particularly to meet the needs of the smart grid. However, even this desktop-client server model is evolving to incorporate web-based editing and cloud technologies. So the heavy client desktop user will be accessing data over the web, even though the desktop users may not even know it. Users will migrate to web-based or smart device technology for simple transactions, like adding an electric service drop to the network.

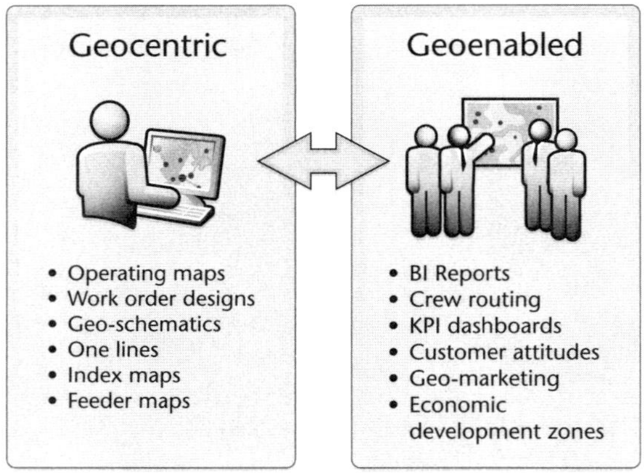

Figure I.4 Geocentric versus geoenabled.

Enormous strides have been made over the last several years to capture accurate land features. Historically, utilities managed their base maps themselves. That will change. Utilities will soon abandon their private base maps and subscribe to a service that publishes up-to-date land information (that is based on cartographic standards) where the data (streets, topographic, imagery) is stored in the cloud.

There will be new technologies to capture data. Despite the decades that utilities have been using GIS, studies have shown that they have only captured a fraction of the real facility data. Utilities have yet to incorporate in their corporate GIS much of the underground structural data or the pipe-type cable data for underground transmission. There will be more use of light detection and ranging (LiDAR), a process that captures points to detect three-dimensional features, and image capture to gather data. Also, there will be new technologies to automate data capture from laser scanning and imagery. Gone will be hordes of people doing field inventory.

There will be a movement toward viewing the facilities model of GIS (the old automated mapping/facility management systems—we will discuss this in Chapter 1) as one component of an enterprise GIS, but not the center of GIS. Instead, this system will be simply a feeder system into an enterprise view of the utility, strongly spatially enabled. Corporate systems, like customer relationship management (CRM), enterprise asset management (EAM), and distribution management systems (DMS), will all be spatially enabled, getting their functionality from web services using GIS technology, probably in a standard interoperability format.

The cloud is the step change. While security and reliability concerns will continue, utilities will figure them out, like they did when they put in SCADA systems. Years ago, people fought tooth and nail to retain staffing of substations, saying that no way would they rely on remote switching over a telephone line that someone could easily hack. The utilities figured it out. Cloud vendors will host the vast majority of enterprise tasks in their cloud. This includes nearly all the applications that involve spatial technology.

Coupled with the cloud will be mobile technology. There has been an explosion of smart phones and tablets, all reliant on the cloud for their intelligence. There will be very little distinction between applications in the office and on the smart phone or tablets. GIS on the smart mobile device will become commonplace soon. Adding data or reporting an event or a problem in the electric system on a smart phone will cause it to show up automatically in the GIS desktop, the web, and on everyone one else's smart phone. The GIS is all about location. Today, nearly every device, smart phones, tablets, and Internet sites are location aware. The GIS will leverage that location awareness.

GIS at utilities are relatively closed to outside data. This will change. The access to online spatial content continues to grow exponentially. A modern GIS will increasingly access and publish data to online communities enhancing communication and collaboration.

Nearly all the computing we do today will be based on some form of web services. Most utilities will move much of their IT infrastructure to a private cloud managed by someone else. Application development time for focused tasks will shrink from months to hours. Application time for more complex applications will shrink from years to weeks. This will be accompanied by a strong self-service trend in utilities, which will rely more on good data from the core GIS. For example, the days of electricians calling in to a call center for a new customer connection will be gone soon. This all will be done with a self-service app on a smart phone or tablet.

The smart grid will change the face of IT in utilities forever. Since the smart grid relies so much on good data from the GIS, this evolving technology will push the GIS to take on a more mission-critical role. The smart grid will require better data management, particularly from the mountains of data that will be coming from smart meters (the so called big data). GIS will do more spatial analysis so utilities can better understand customer behavior in relationship to their delivery infrastructure. The smart grid will demand solid data management, spatial analysis, and better visualization and dissemination of the information from their GIS.

So going forward, there will be a need for a number of desktop users of GIS creating new content. There will be a strong place for ruggedized devices in the field for the hardcore workers, but they will probably be more web enabled. However, there will be a greater need for servers and subscription services for the rest of the information in the cloud accessed through consumer-friendly mobile devices.

CHAPTER 1

GIS and the Business of the Utility

Electric utilities face enormous challenges. They face increasing:

- *Regulatory pressure:* including meeting demands to reduce carbon emissions and add renewable resources;
- *Demand:* despite the financial downturn demand will surely rise, especially as the transportation sector moves away from petroleum and toward electricity;
- *Infrastructure age:* as infrastructure is aging faster than it is being replaced;
- *Worker age:* as utility workers represent one of the oldest workforces in industry;
- *Customer demands:* as customers are used to instant gratification and online access to information, and they are far less tolerant of interruptions;
- *Costs:* despite the economic downturn costs continue to rise, particularly labor costs;
- *Transparency:* as utilities are easy targets.

At the same time the resources available to the electric companies are shrinking. They face decreasing:

- Revenues;
- Fuel options;
- Tolerance for mistakes;
- Access to capital;
- Public acceptance of rate increases;
- Land resources to build their infrastructure.

Utilities will need to change quickly to meet these pressing challenges. They will need to rethink many of the ways they have been doing business for more than a century.

A GIS will not keep older workers from retiring, make access to capital easier, or lower the expectations of customers. It will not stop wind farms from killing birds or make solar energy more affordable. However, a GIS will provide a spatial context for problem solving. The vast majority of problems facing power companies involve location, such as:

- Where to site a wind farm;
- Which transmission lines are at a high risk of failure;
- Where to search for geothermal resources;
- How to run a transmission line from a proposed power plant to the grid;
- Where to place sensors on the electric transmission and distribution system to create a smart grid;
- Where to run telecommunications systems to best communicate with smart meters;
- Which customers hate their trees trimmed;
- Whether there is a spatial correlation between customer satisfaction and high bill complaints;
- Where there are sagging transmission lines out of compliance;
- Which substations will be under water during a 100-year storm event;
- Which transmission lines are located in tornado prone areas;
- What the impact is of a rising water level of coastal electric equipment.

GIS: A Spatial Context for Solutions

GIS helps solve problems by making it simpler to discover relationships.

The problems listed here are tough. Since GIS is such an integrating framework, and since the common denominator or theme of these questions is location, utilities can display a variety of data sources and results of analysis together in the same view. That consolidated view is often called a common operating picture or situational awareness dashboard like shown in Figure 1.1.

GIS is more than just seeing information on a map. It provides the ability to see relationships and help determine not only the "where" of something bad happening, but provide the spatial context of the "why" something might occur or has occurred. It extends the data to give answers to problems that otherwise would not have been seen. It connects the dots.

Transforming a Century-Old Industry

Utilities can apply GIS to many of the workflows in the various departments of the power company. Electric utilities have been around for over a century. Like most mature industries, work processes became compartmentalized. That's a polite way of saying that departments and organizations operate in their own silos. As information systems developed over the last several decades, the information systems supported the silos. Sharing data from one organization to the next became more and more difficult. As utility operations grew increasingly complex, data duplication became commonplace.

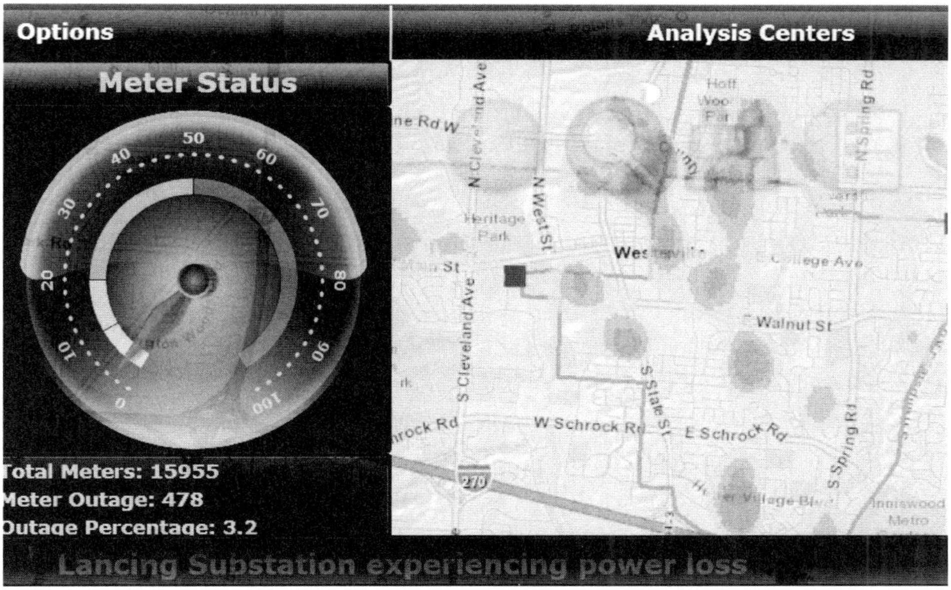

Figure 1.1 A GIS common operating picture consolidates data from different sources (Source: Esri et al.)

Do You Know Where Your Poles Are?

One utility in the Southwest United States reported that data about a wood pole existed on as many as 40 records—some stored on spreadsheets, others in databases, and even some still in notebooks. The information about poles was stored in so many different places because each business unit that dealt with poles kept their own records of pole information, either not trusting another business unit or more likely not even aware there was data duplication. That same utility did not have an accurate count of their poles. The accounting systems reported one number, while the engineering records had another. Since this utility and most utilities in the U.S. have to pay property taxes for their assets, they had no clear picture of what they should be paying in taxes. This same utility also had to report to their public utilities commission (PUC) their total value of their installed plant on a regular basis in a rate case. Should either the local municipalities or the state PUC call for an audit, there was a good chance that the value of their pole plant would be incorrect. An audit would likely result in some kind of fine, back payment, or at the very minimum public embarrassment.

Bad things happen when a utility cannot coordinate its data. For example, if cable inspection data is not coordinated with age or capacity data, operators might run the cables too hot and they may fail. If up-to-date flood data is not coordinated with substation location, then substations could end up under water during a big storm.

Utilities deal with data in a variety of forms. In fact, some of the data that utilities use to run their business is not stored anywhere. Older workers pass vital information from their generation of workers to the next. Much of the data the utility uses, perhaps as much as 90 percent, has some degree of location associated with it. Where should a new transmission line run; what is the evacuation limits of a nuclear plant; where do the emergency workers live; where is the current location of a trouble truck; where are insulators coated with salt; where are lightning arrestors that were recently recalled due to a manufacturing flaw? Location is the common denominator for much of the disparate data a utility uses (and much of this lives only in the minds of its workers).

GIS Helps to Put Locational Data in Order

A modern GIS has three main capabilities:

- Managing data;
- Performing spatial analysis;
- Creating awareness.

Managing Data

The first capability of the GIS is data management. Regardless of the format of the digital locational data, the GIS has the ability to accept information as long as there is some kind of spatial reference, such as latitude/longitude or street address, and to organize the information in the form of a map. It can then create layers of one kind of spatial information displayed over another. If a utility has a customer satisfaction problem, the utility could create one layer that is the result of a customer satisfaction survey with the average scores (for example, from 1 to 10) organized by zip code. Another layer could be average distribution reliability by zip code, measured say in average annual customer minutes out of power. Together they tell a story about which areas have poor reliability and low customer satisfaction. Then if the utility wanted to gain more insight, the utility could add a third layer that lists average income by zip code. With these three spatially coordinated layers, the utility could gain ever-greater insight into the how certain groups within income levels view reliability over other groups. The utility could add other layers such as where there is a concentration of high bill complaints. The GIS has the ability to organize these various layers, even if they have different projections and accuracies and simply visualize the layers together.

This notion of organizing layers of information together is not new. People used transparencies to illustrate how one set of images relates to another. Before GIS, planners created overlays that displayed street networks on one transparent layer. They then created another overlay, say, zoning boundaries. They created other overlays, such as vegetated wetland boundaries. Another layer might be vacant land or conservation land. They lined up the various overlays along some kind of reference, like grid lines and could then visualize all layers together. The purpose of combining these overlays was to gain a new understanding of the data. So these layers could show the city planner which areas could or could not be developed. The purpose of combining these layers together was for decision making and action.

The GIS does this process of combining different types of data digitally for the same purpose—to gain additional insight as to what action to take or what policies to develop.

Spatial Analysis

Just seeing the various layers is interesting, but the real power is to perform analysis on the layers. This is the second capability of the GIS, spatial analysis. In the previous example, suppose the city planner wanted to find a parcel that was zoned industrial, was not in a vegetated wetland, and was within a mile of a major highway. Depending upon how large the area of study was, in the old days using the transparencies that process could take considerable time. The planner would have to study each parcel, measure how far each possible parcel is from a highway and then mark the parcels that met the planner's qualifications. Using the GIS's data management capabilities, that process would be easier, since the planner could easily pan and zoom around the display, turning on and off layers, but it would still be tedious. Using spatial analysis, the GIS could perform a spatial regression, along with a distance optimization, and discover those parcels within a mile of a highway, zoned industrial, and not in a vegetated wetland. This process would take seconds.

Awareness

Once the GIS performs the analysis, the planner would like to show the results in a way that is meaningful to others. The notion of awareness is the third capability of GIS. In the previous example, once the GIS determined which parcels met the criteria, the planner now needs a means to visualize and disseminate the results. This allows others to gain awareness of the analysis and to be able to take some kind of action. The GIS can package the various data layers (the zoning, the wetlands, the parcels) along with a new layer, which the GIS creates, that shows only those parcels that meet the criteria and publish the results to all interested parties. Today those results can be published to the web, desktops, and mobile devices (e.g., laptops, tablets, and smart phones).

The three essential capabilities of the GIS are as follows:

1. *Data management*—the ability to consume virtually any kind of data that has some spatial reference and organize the data along some common coordinate system;
2. *Spatial analysis*—the ability to transform the spatial data (in whatever form) into new information using a wide variety of spatial operators, like intersections, unions, clusters, buffers, measures, regressions, and hundreds of other operators to produce new information that often is the answer to a business problem;
3. *Awareness*—the ability to visualize the collection of data as well as the results of spatial analysis in the form of a map (or a report, chart, or diagram organized by location) that facilitates groups sharing that information using a variety of delivery mechanisms (e.g., the web, a smart phone, a desktop).

In the previous example, the city planner used the GIS to collect data from a variety of sources, perform a spatial analysis on the data to produce a result, and publish the results to the city officials, other departments, and maybe even the general public. See Figure 1.2 for a diagram of the three GIS capabilities and how they interrelate.

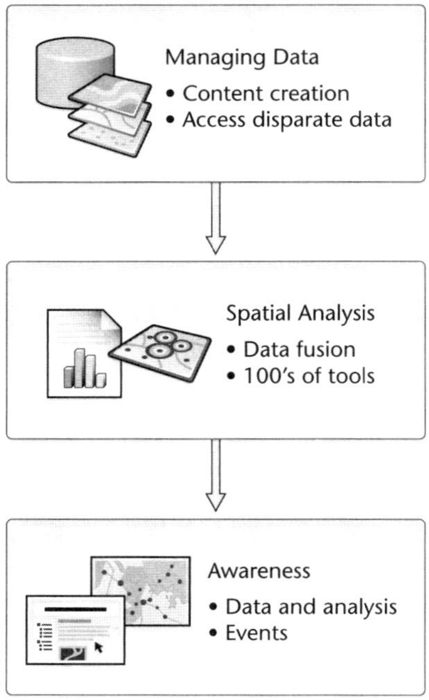

Figure 1.2 GIS capabilities.

Development of GIS at Electric Utilities

To fully understand where GIS needs to go, it's helpful to understand how GIS has developed at electric companies. GIS has its roots in the mapping departments. Utilities have used maps since the beginning of the electric utility industry. As they rolled out their new facilities, they needed a means to keep track of what was built in the streets, alleyways, and under the ground. The most convenient way was to show the poles, wires, pipes, and transformers on a map. Some utilities were better at this than others. Some just captured the medium voltages; some took careful measure of the phase relationships; some even detailed the type of open wire pole construction, such as that shown in Figure 1.3. Others were not so careful or complete. While industry groups issued standard symbols for electric power equipment, rarely did these standardized symbols find their way onto the mapping systems.

One other aspect of the old hand-drawn mapping systems was the lack of adherence to any cartographic or coordinate standard. Utilities either created their own street maps or used whatever available map books happened to be published at the time. The notion of the map book became the standard format for the utilities. In the early days of the utilities, map books were organized into pages and each page represented a piece of geography. Often across the top of the page was a series of letters, and along the side a series of numbers. In the back of the book, the map book publishers created an index. They assigned each street in the index to a page in the map book and a letter and number. This created a map grid, with each grid having some kind of identification. Utilities adapted this scheme to their early mapping systems. Often they made each grid into a separate map sheet. They

Figure 1.3 Typical overhead pole construction. (Photo by Bill Meehan.)

then drafted onto to this map sheet the electrical facilities that were contained within that map grid. Over the years, these map grid numbers found their way into other information systems, like customer information systems, and outage and work management systems.

Mapping Systems Get Dysfunctional

These map sheets got more crowded as the electric system developed, so utilities created additional map products, such as overview maps and index maps. Sometimes, they created one set of maps for different systems, like one for the medium-voltage network and another for the low-voltage network. One utility created a completely different series of maps for their secondary network system, radial underground system, and overhead primary system. Some separated their map products by voltage level. What happened, of course, was that these mapping systems became increasingly unwieldy. Utilities found that it became more and more difficult to keep these maps up to date. Changes came rapidly during periods of high expansion. Operating groups grew concerned about the currency of these maps and would keep their own separate maps, often not sharing their maps with the mapping department.

Old Mapping Workflows

Whenever the utility received a request for a new service or the engineers determined the distribution system needed to be upgraded, they created a work order or job order. The work orders started out as a description of the work to be done, like adding a new extension to an existing distribution line or installing a new transformer to feed a new shopping mall. Designers created construction documents that illustrated to the line crews exactly what they needed to build. They included sketches showing construction details. Almost always, these construction documents included some kind of map, showing the location of the work to be done and the details of where they crews needed to install new equipment.

The basis for these maps were the hand-drawn operating maps that illustrated the current location of the existing system, the so called as-is condition of the electrical system. Once the crews completed the work detailed in the work order, the line foreman or supervisor marked up the construction drawings, including the maps, to show what actually had been built compared to what the designers had specified. Often what the designers actually specified was somewhat different from what the crews actually built in the field. These marked-up sketches then were returned to the engineering department. There the designers or mappers would take the marked-up sketches, called as-built sketches, and incorporate this information onto to the various map sheets. Since these map sheets were hand drawn, the process of updating these maps was extremely time consuming.

It was not uncommon for a utility to accumulate a large backlog of these marked-up work order sketches. As the electric systems became more complicated and extensive, utilities needed to add more and more people to the mapping department just to keep up with the backlog of data. If the utility allowed the backlog to grow too high, the engineers, designers, and field operating people ended up using information that was out of date. This created all kinds of problems and costs to the utility. Crews would be dispatched to wrong locations or customers would be out of power longer, since the trouble shooters, who searched for the cause of a failure, would not be using the most up-to-date information to determine the location of a problem.

Development of Digital Mapping Systems—the Birth of AM/FM

Thus, utilities needed a faster way of updating the operating maps from the as-built sketches created by the field. Computer graphics developed during the late 1970s. Some utility early adopters converted their old hand-drawn maps to digital maps. These early systems continued to develop well into the 1980s. By the mid 1980s, digital utility mapping systems had grown popular. Utilities began to demand that these mapping systems do more than just make editing maps faster, but saw a need to somehow link attributes of the various equipment shown on the map to the cartographic representations of the features. These early utility mapping systems were called automated mapping/facilities management (AM/FM) systems.

The term AM/FM still is used today, although it is often used interchangeably with GIS. Often, the term AM/FM/GIS is used to describe a digital mapping system that is primarily used to document features of the electric system (most often the electric distribution system). For many utilities, the term GIS implies the same capabilities of an AM/FM system. While AM/FM systems helped the problems associated with the constant editing of the operating maps, their use was limited to the operations and engineering department. In effect the information products produced by these systems were essentially the same as the information products produced by hand-drawn maps, with the exception that it was easier to create some statistics (e.g., how many transformers were in a certain map grid). It was also easier to locate equipment by name using simple queries.

The focus of AM/FM systems was (as the term implies) to create a digital operating map (that looked almost exactly like the manual maps) and to document the facilities that made up the medium-voltage and sometimes low-voltage distribution systems. However, utilities soon discovered that the promise of using these systems for other business needs or for additional information products was sorely lacking. When utilities began to see the value of combining their trouble call systems with digital mapping, they saw that critical information was either missing or inaccurate. In the U.S., for example, much of the medium-voltage system consists of a mix of three-phase and single-phase lines. Yet in a large number of cases, phase determination was missing. In some cases, the connectivity of the lines was either missing or not precise. Thus, the value of using the AM/FM system as a foundation for outage management was less than they had hoped for.

Given the amount of data and the redundancy of the data sources themselves, just on the distribution system, the cost to convert this data to digital form was enormous. The idea of an AM/FM system was to replace the very labor-intensive hand-drawn mapping process and to reduce the steadily climbing backlog of work orders. However, utilities often failed to capture any additional information (like phasing or connectivity or underground structures) that they might need in the future. The result was that the vast majority of AM/FM systems were largely used only by the mapping department. The output products were cleaner and easier to read, but were no more informative than the hand-drawn maps that proceeded them.

Not Much Changed

The consumers of the AM/FM systems didn't change their processes either. The mapping group plotted their computerized maps on to transparent paper. Just as they had been doing for decades, the mapping group copied the fresh plots using the old blueprint machine and mailed the prints all around the company. Other than the print quality, the consumers of the maps hardly knew the difference. In fact, whenever supervisors, dispatchers, and other field workers noticed discrepancies from the prints, they marked up these copies with red markers. They were reluctant to replace the newly copied revisions in their files, because someone would have to check the new print against the old marked up print to see if all the new print contained all the corrections from the old print. Too often, the new prints would pile up just waiting to be reconciled with the prints the field people used. This workflow was identical whether the AM/FM system created the new print or the print came from an old hand-drawn map. Figure 1.4 illustrates the workflow and how awkward it was.

After spending the millions of dollars converting data from the hand-drawn maps and records to the AM/FM system, the utilities didn't really solve the problem of increasing backlog of work, out-of-date maps, and inconsistencies between the mapping system and the information about what was really going on in the field. Utilities rarely worried about the absolute accuracy of the location of what they depicted on the maps, as long as they were relatively correct. Many made sure that their old mapping grids and identification systems remained. These digital mapping systems were good at replicating the old mapping products and for some accounting and reporting data. Rarely did they have the ability to do any kind of spatial analysis.

While utilities used computers to generate the maps, the concept of the map sheet, a finite piece of paper, with specific dimensions remained intact.

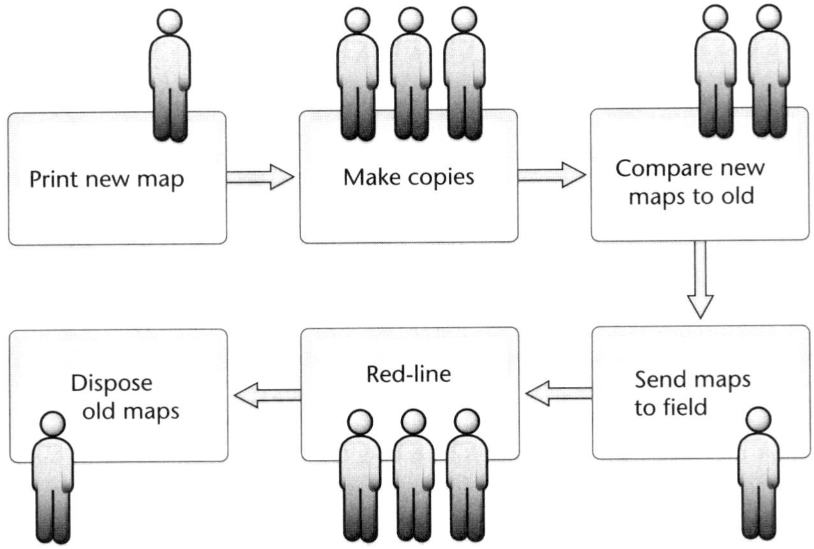

Figure 1.4 Common workflow for updating maps in the field.

GIS Is Different

At about the same time that AM/FM systems were being developed, GIS in its truest form was rapidly developing. GIS, unlike AM/FM, was about solving spatial problems, not just about documenting assets. One of the earliest applications of GIS in the utility field was electric transmission site selection. The problem facing a transmission operator is to find the best path for a new transmission line subject to a number of constraints (See Foreword.) Prior to GIS, transmission planners had to analyze different factors, such as slope, earthquake zones, proximity to population centers, environmental concerns, wildlife impacts, and access. The process involved gathering data from many different sources, then attempting to piece them together into some kind of rational map to help the designers come up with the lowest construction cost and risk proposal, with the least amount of permitting and public relations problems. These assessments could take months. What often happened then was that during the permitting process, citizen groups and other interveners could block the progress of the project. The result was that the utility would in effect go back to the drawing board and begin the route selection process all over again.

The idea of GIS was to create a way to capture geographic information in a database. Information such as the topography of the land, locations of problematic areas, indigenous population burial grounds, sensitive habitat areas, or areas that were sensitive from an aesthetic perspective could be stored as separate GIS data sets or layers. Then, once the GIS normalized these factors, it displayed the results on a map. The map could show a variety of corridors, color coded by cost or risk. Once the GIS analyst builds the model gathered by the data, the planners could meet with the public and government officials to present their proposals. If things needed to be changed or some new factor came to light, the model could be run again quickly and the new plans created. With GIS, tweaking the model was simple. Of course, in the early days of GIS, the tools were less rigorous and harder to use, and computing power was substantially weaker.

The maps produced by GIS contained new information, never before seen. The maps produced by AM/FM systems documented what people already knew.

Networks Are Added to GIS

Where GIS was weak was in the modeling of linear networks, like electric lines. However, during the development of GIS, planners discovered that they could apply the same concept of discovery to linear networks. The idea of adding linear network data into a relational database along with the other spatial data helped solve routing problems—not just the routing of transmission lines, but of vehicles and people. Like spatial analysis with GIS, the result of the linear spatial analysis was a map showing new information. Once GIS modeled transportation networks, by the mid 1990s, most GISs were able to model linear networks. Utility networks, like water, gas, electric, and telecommunication systems, now could be modeled and more importantly analyzed using a GIS.

Once utilities were able to model networks in the GIS, they began to abandon the old AM/FM systems. However, while many utilities have migrated from their legacy AM/FM system to GIS, the problems of missing and inconsistent data due to the lack of a standard coordinate system that the GIS inherited from the AM/FM limits its use for those utilities. For some, even though they migrated from their old AM/FM systems to GIS, they continued to use the GIS in the same way.

What's the Difference Between CAD and GIS?

During this same time period, computer-aided design (CAD) systems developed and matured. Since CAD can produce any kind of drawing, some utilities had opted to use CAD to capture their mapping information, rather than to adopt an AM/FM system or GIS. Utilities installed CAD systems in their design and drafting departments, migrating from hand-drawn detail design drawings to CAD. It is still common and appropriate that most engineering departments today use CAD tools to design new substation, vaults, transmission towers, structural members, and power plants. The result of the design process is a set of detailed construction drawings.

The design departments had the same problems as the mapping departments. It was hard and time consuming to make changes to hand-drawn plans. Early CAD systems were to drafting what word processing was to typing. Hand-drawn design drawings were very difficult to create and more difficult to edit. If a drafters needed to move a detail or even a piece of text on a manual drawing, they would have to erase large portions of the drawing only to have to redraw the same thing in different place on the drawing.

With CAD, like the word processer, the drafting process was enormously simplified and automated. Drawings were much faster to create from scratch and simple to edit. They were also easier to read. They could be stored digitally, eliminating the costly drawing vaults. CAD systems developed further to automate the design process itself. CAD systems so simplified the drafting process. Since CAD worked so well designing structural steel details for substations, why not use the same systems for the mapping departments? Many did. So those utilities simply digitized their old circuit maps using CAD tools.

What utilities discovered was that while CAD automated the drafting process, it was not able to perform the tasks of data management, spatial analysis, and spatial awareness that utilities needed. CAD treated the maps as a series of discrete drawings, not as a single information system. Typically each map was stored as a separate file, as if the distribution system consisted of a series of separated pieces within a map grid. The boundaries of an old map grid do not limit the GIS. It doesn't store each map sheet (as they had been originally drawn manually) as a separate file. Instead the GIS provides new information—it does not just replicate old map sheets in an automated way.

CAD, of course, is an important tool in an electric company. It is critical for creating complex designs of generating stations, for detailing projects, for providing structural details, and for developing construction plans. The purpose of a design drawing is straightforward: utilities use it once to document what a contractor has to build. For example, once the contractor embeds the rebar into a concrete foundation, there is very little need for the rebar detail design drawings.

CAD is not a GIS. Once built, the utility needs the GIS to continuously manage the utility.

As noted before, GIS is a spatial data management system, built on top of a database management system. CAD is project based, typically file based, as each project or design drawing is created in a separate file. Can a utility use CAD for its GIS needs? Yes, with some work, compromises, and probably a significant amount of customization. Can a utility use GIS to create detail substation designs? Yes, with some work and a lot of compromises as well. A spreadsheet can certainly be used for word processing, but it would be much easier to use a word processer. Can a word processing program be used for balancing rows and columns of numbers? Yes, but not nearly as effectively as a spreadsheet. CAD has its place and GIS has its place.

GIS Architectures

Computerized mapping started as mainframe or mini-computer systems. They evolved into standard client server systems. Users worked on desktop PCs in a thick client environment. Data was stored in a proprietary or commercial relationship database system on a database server. Integration with other corporate systems like supervisory control and data acquisition systems (SCADA), customer systems, and work management systems were largely custom built one-off interfaces. More often, data from one system was extracted from corporate systems then loaded into the GIS. Early uses of GIS in the field mainly consisted of extracted information from the GIS loaded into thick desktop-like mobile clients. See the typical "classic" configuration in Figure 1.5.

GIS Is Changing Rapidly

Today, this architecture of GIS at electric companies is largely unchanged. Integration has improved and utilities are using thinner client mobile technology. This will change dramatically soon, as utilities fully leverage the cloud and as smart de-

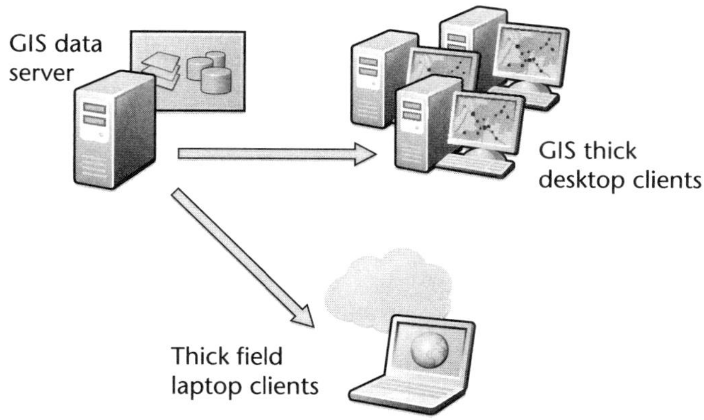

Figure 1.5 Historic GIS architecture.

vices become the dominant mobile devices of employees. Also utilities are adopting service-oriented architecture and web services more commonly.

Figure 1.6 shows a more modern approach to GIS, one in which some of the spatial data is not contained with a single GIS, but found using web services. That web services might be from the open Internet or a subscribed service or some combination. Cadastral information, imagery, and parcel information may come from completely different sources and be consumed within the electric company GIS. The real estate department, for example, could maintain its own records of land ownership and easement records but publish that data through the company's intranet, and it would or could be consumed by a spatial analytic model in the system planning department's view of the GIS.

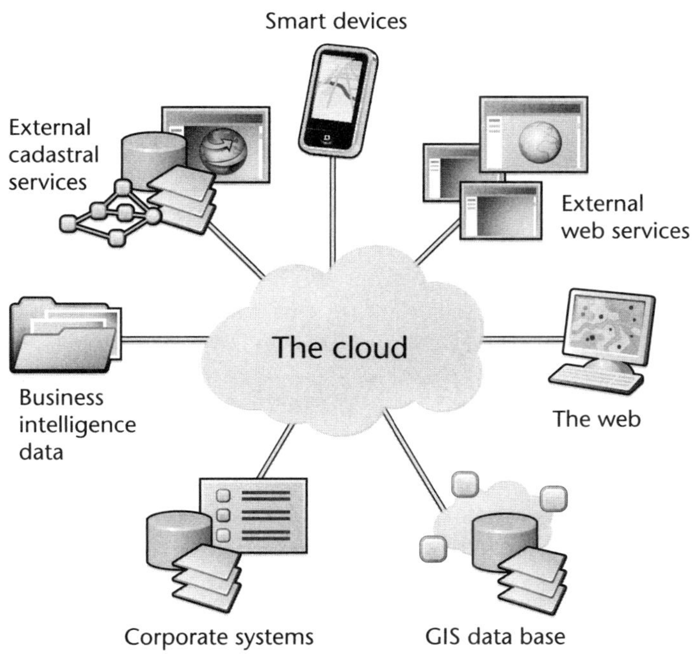

Figure 1.6 Next generation of GIS architecture.

What Matters Most to the Electric Utility

GIS is growing up and growing in its importance in the electric company. Today most utility people recognize that GIS is essential for any reasonable smart grid deployment.

The purpose of the GIS is to improve the work of the utility not to automate the mapping process, like in the old AM/FM days. It should positively impact what matters most to the company. So what really matters most to an electric company? The preface described what one savvy power company CEO thought:

- Make money for the shareholders;
- Keep customers happy;
- Keep employees safe;
- Obey all laws and regulations;
- Keep the lights on;
- Respect the environment;
- Keep us out of trouble and out of the newspapers.

The Balanced Scorecard

Examining this list, there emerge four stakeholders that the utility must serve:

1. *Shareholder*—the owners of the company, either the actual stockholders in the case of an investor-owned utility or the government (city, regional, state, federal, tribal) or members of an electric cooperative;
2. *Customer*—the end use customer of the electric service;
3. *Employee*—actual employees, contractors, and vendors;
4. *Community*—the broad group of people who make up the service territory, including the representatives of the community such as the regulators (public utility commissions, ministries of electricity, environmental agencies), government officials, special interest groups, and perhaps even the media.

Satisfy those stakeholders and the company accomplishes its mission. There are just a handful of strategic ways to satisfy these four stakeholders. To fully understand if the GIS is having a positive impact on the company, it is helpful to articulate the mission in terms of key performance indicators (KPIs) or metrics for each of the stakeholders and then to measure the impact of the GIS on each of them. For example, one shareholder KPI for an investor-owned utility is earnings per share. This is essentially the profit of the company divided by the number of shares of common stock. Profit depends on revenue and expenses. So the GIS must show that it contributes to the improvements in productivity of the workers, reductions in travel time, savings in energy, effectiveness in planning for capital projects, and really anything the utility does that involves spending. The GIS could help to improve revenues by ferreting out places where theft of current occurs or helping collectors do their job to increase delinquent payments.

An example of a KPI for the customer stakeholder is "keep the lights on" or, to be a bit more precise, maintain reliability. Electric companies most often measure reliability using standard measures. One popular measure is system average interruption duration index (SAIDI), which is a measure of the number of minutes on average a customer experiences an outage annually. This is also referred to as customer minutes out. The GIS should have a positive impact on a utility's SAIDI. Other important customer KPIs are the customer connect time—the time from when a customer requests a new electric service to when the utility connects the service and energizes the customer.

Electricity is dangerous. Utility must keep their employees safe. Employee safety is almost always one of the key performance indicators for a utility. Utilities measure safety performance as lost time accidents and responsible motor vehicle accidents. Other important measures are switching errors or near misses. Good facility data, accurate location, and easy accessibility to spatial data, served from the GIS, are critical safety nets that contribute to employee safety.

Nearly all electric utilities worldwide are regulated. Regulators are the representatives of the communities that the electric company serves. One important KPI of the electric company is the number of fines that the regulators impose. These fines could be for poor emergency response or late communication over excessive emission release or for a utility truck driving over a protected wetland. These violations are expensive, but more importantly they erode the position of the company in the community. In Russ's words (from the Preface), these are the things that get us into trouble. The GIS is a critical tool for helping utilities better plan for emergencies and mitigate risk.

Often, it is a difficult task to keep all these stakeholders happy. One way to do this that utilities have found useful is through the use of a balanced scorecard. The scorecard is a series of high-level KPIs organized by the four stakeholders. Figure 1.7 lists an example of a typical balanced scorecard for a transmission and distribution utility.

GIS Helps Visualize What to Balance

The list of KPIs in the balanced scorecard is often an articulation of the company's mission as well. Most electric companies strive to return earnings to their owners,

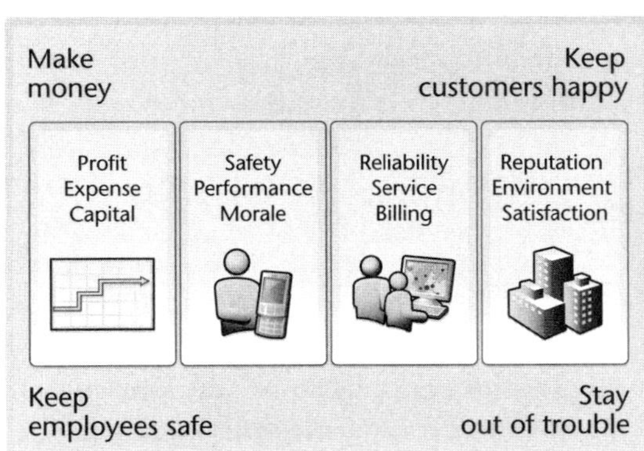

Figure 1.7 Key performance indicators in the balanced scorecard.

satisfy their customers with good service and reliability, provide a safe working environment, and be a good neighbor. When the executives gather to plan for the upcoming budget year, they will often set goals around these critical KPIs, like lowering operating and maintenance costs, increasing earnings, lowering SAIDI by a number of minutes, setting aggressive safety targets, and reductions in community issues. The gap between current performance and future targets represent the plan of action of the year as well as funding to achieve these targets. Most often, they close these gaps by improving upon processes, which almost universally involve the application of information technology.

Since nearly all of the KPIs of an electric company involve some element of location, the GIS should play a pivotal role in the gap reduction to achieve the desired performance levels. However, given the GIS's more traditional role as a mapping system or an engineering system, it has not had as dramatic an impact on the critical KPIs as it could have. One of the goals of this book is to help point out areas where the GIS can in fact be deployed to help close those gaps in the key performance areas.

What an Electric Company Does

As noted previously, what matters most to electric companies is articulated in their mission statement and is measured in their key performance indicators. There are thousands of workflows that contribute to a single KPI, from repairing a broken insulator to exercising a valve in a power plant to rendering a pay check to an employee. Most are rooted in location.

Workflows are organized around these themes:

- *Managing performance*—these workflows center around keeping the system running smoothly and efficiently;
- *Managing the infrastructure*—these workflows focus on the ongoing development of the electric infrastructure and its ability to continue to meet the performance of the company;
- *Managing support services*—these workflows support the ongoing workflows of managing performance and managing infrastructure;
- *Managing information*—these workflows contribute to the knowledge the company needs to make decisions about performance and managing the infrastructure.

Figure 1.8 illustrates the four strategic processes of the utility.

Managing Performance

Performance separates good utilities from not so good ones. Power plants must continue to operate, the lights need to stay on, customers need to be able to get service, and bills have to be accurate and rendered on time. In other to sustain the business, utilities must make profit or, for government-run utilities, not lose money. Regulators, investors, and boards of directors generally measure utilities on how

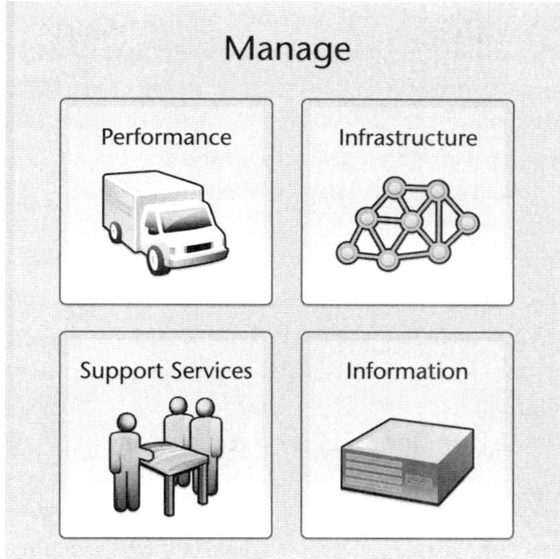

Figure 1.8 What the electric company does.

well they perform, not so much on how hard they work. Performance workflows influence all four of the stakeholders.

Common workflows that strongly influence performance are inspections, preventative repairs, tuning, dispatching, customer care services such as billing and account assistance, answering customers inquiries, monitoring the system, and emergency management. Assuring cyber and physical security, collecting money from delinquent accounts, keeping inventory levels low, and calculating and mitigating risk are all examples of performance workflows. All have a strong locational component, and all are driven by the need for good spatial information and the ability to analyze that information.

Metrics that contribute to shareholder value include things like heat rate of a power plant, number of breaches of physical and cyber security, bad debt, electrical losses, theft of current, and estimated meter reads. Metrics that impact the customer are the reliability statistics like SAIDI, billing complaints, voltage dip, and low-voltage complaints. Metrics that impact the employee are accident frequency, near misses, employee morale, and employee turnover. Finally, metrics that impact the overall community include number of hazardous material accidents, fines for environmental violations, customer satisfaction, complaints, and negative (or positive) social networking mentions.

Utilities categorized spending for these activities as operations and maintenance (O&M) expenses. O&M expenses directly impact the bottom line and earnings. Effective (or ineffective) management of performance directly impacts customer service, safety, community relations, and of course the ability to make money. The key for GIS managers is to clearly articulate the improvement in performance metrics directly attributable to the GIS.

One simple example, which will be covered in more detail in Chapter 8, is in the deployment of field work crews. A large number of employees work in the field. So, part of their work day is consumed by traveling from one work site to another. Many utilities do not take advantage of the capabilities of the GIS to optimally dispatch crews. This goes well beyond finding the shortest direction from one point to another. This means taking the entire field group and determining the best use of all of the workers in the work group to minimize travel time. Reduction in travel time to perform the same series of tasks lowers labor expenses, saves fuel (and carbon emissions), and improves safety, since the crews are driving less.

Manage the Infrastructure

In order to keep the performance at a high level, utilities must continue to develop their infrastructure to meet changing future demands. Ineffective projections of their customer demands and subsequent failure to upgrade their infrastructure could leave the system vulnerable to overloads, outages, and brownouts. Common workflows are system planning, load forecasting, work order taking, asset management, engineering and design, site selection, construction, as-built documentation, provisioning for new service, obtaining easements for new facilities, and a host of other activities that support new infrastructure development. Again, these workflows depend heavily on location and of course good data about the current state of the infrastructure.

Measures that impact the shareholder are on time project performance, asset utilization, change orders as a percentage of project cost, construction variance from design, backlog of work orders, and closeout of construction documentation to the plant accounting records. Metrics that impact the customer are meeting customer in-service requirements, meeting interconnection schedules, and insufficient energy to meet the demand. Metrics dealing with employees for managing infrastructure are similar to those for managing performance and include project safety, union grievances, morale, and turnover. Metrics dealing with the community are similar for both managing performance and managing the infrastructure and include complaints of noise, disruption of wildlife, complaints, and fines for contamination during construction.

Utilities amortize the spending for these workflows. This is capitalized (i.e., amortized over the life of the new assets being added to the system). The value of the assets includes all costs of the development of the new assets from planning costs, engineering, data development, and acquisition to labor, materials, and all administration of these activities. These costs indirectly hit the bottom line, since only the carrying costs directly impact earnings but they impact the total debt of the company. The value of assets is regularly a factor in determining the cost of service for a regulated utility, which in turn determines the rates the utility can charge, subject to regulatory approval. Transparency of these costs is critical for a successful rate case. (A rate case is a regulatory filing to the public utility commission, ministry of energy, or some other regulatory agency often held every three or four years to determine the rates a utility can charge.)

Graphic design of new distribution facilities and as-built distribution network documentation are a few of the earliest applications of GIS in electric companies that deal with managing the infrastructure. These applications help the utility accurately document what assets they added during a specific time period. Prior to GIS, utilities kept rooms full of asset information. During the rate case preparation, they needed to hire hordes of people to assemble the documentation of the costs and time frames of what was added where. The utilities financial viability depends on knowing exactly what and where their assets are.

Managing Supporting Resources

Support services include all of the workflows that support either ongoing performance or managing the infrastructure. It's in these areas that GIS has not played a significant role. However, much of the work of these support services, sometimes called shared services, has a strong spatial context.

Shared services workflows and application of GIS will be detailed in Chapter 8. Examples of these workflows are fleet management; supply chain (purchasing, warehousing, materials management, and delivery); and providing services such as legal, human resources, financial management, accounting, auditing, labor relations, public relations, real estate management, and investor management. Yet, some of these areas represent where the company's biggest problems may lie. For example, utilities write off bad debt each year. Once the debt is written off, it can longer be collected. They, in effect, sell off the receivables to collection agencies for pennies on the dollar. Utilities could use GIS to help them focus their collection activities in those areas where collection activity is most likely, such as in high-income areas. Instead, they can get overwhelmed with the sheer volume of customers who do not pay their bills.

Managing Information

Today data can inundate utilities. Workflows include the maintenance and operation of the IT infrastructure, managing services such as backup and security, application development and deployment, data acquisition, planning, support, training, and network operations and often includes the management of the telecommunications networks, which are becoming more and more complex.

Metrics relate back to how well IT is supporting the overall mission. GIS of course is one of the critical components of IT, but it is also or can be an important tool to help manage the assets of IT and help with the emerging concerns of physical and cyber security of IT assets. In order to leverage that spatial component of the data, it's critical that utilities incorporate their GIS with their other major corporate systems, like their financial, material management, network analysis, asset and work management systems, and their smart grid support systems such as SCADA, distribution management systems (DMS), and meter data management systems (MDM).

The Utility Value Chain and Its Spatial Nature

Utilities are not organized along the lines of the managing performance, managing infrastructure, managing support services, and managing information. Rather they are organized along functional, business, or process lines. Electric utilities tend to organize themselves are around the energy delivery value chain. For much of the twentieth century, utilities were vertically integrated, meaning they managed all aspects of electric production, delivery, and the sale to the end user. They built and ran power plants, built and operated high-voltage lines, built and managed distribution systems and substations, read the meters, sent out the bills, and collected the cash. They often had their own testing labs and, in some cases, even manufactured their own equipment. Today, that is less common.

Location Critical for Energy Supply

The energy supply value chain consists of these master processes. Note that within each of the master processes, the operators manage performance, infrastructure, support, and information.

- *Exploration*—this is the process in which the actual source of the energy is uncovered. This process is mostly performed by the fuel providers, but today utilities (or subsidiaries of the utility) are often involved in the discovery and validation of a fuel source or a commercially viable fuel source for electricity production. Today those sources are wind, solar, biofuels, and geothermal.
- *Development*—once the fuel source is determined and validated, a production facility is developed.
- *Gathering*—in the cases of, say, a wind or solar farm that consists of many smaller production units, there exists a gathering system to collect the power.
- *Production*—gathered energy is packaged for delivery or a power plant produces the energy.

The energy supply organization handles these processes. How GIS enhances the energy supply organization will be detailed in Chapter 2. In addition to the actual production, energy supply organizations also buy and sell power, issue power contracts, look at their financial risks, and hedge their purchases. Even though many utilities have divested of their generating assets, most electric companies still have to manage their energy supply portfolio. Location plays a strong role in the energy supply business, from discovery of new resources to site suitability studies to load forecasting based on weather and historic trends. Once they build plants or wind farms, GIS can play a strong role in managing these complex facilities.

GIS Plays a Strong Role in Transmission

The next process in the value chain is wholesale delivery or transmission. GIS plays a strong role in modeling the actual transmission facilities, transmission routing, vegetation management along the right of way, dealing with difficult land ownership, encroachment concerns about electromagnetic fields (EMF), aesthetics, and lightning protection. There is a high degree of control and coordination that exists between production (energy supply) and transmission. GIS has been used to some extent in transmission, although rarely has it been leveraged beyond the ability to produce transmission strip plans or facility management plans. Transmission substations are rarely included in the GIS, but provide a critical component of the transmission system. Without including the transmission substations in the GIS, the ability to truly model the behavior of the transmission system is compromised. Chapter 3 deals with how GIS enhances the performance of electric transmission.

GIS Enhances Distribution

Utilities first used GIS to model their distribution assets as noted earlier as an outgrowth of their mapping systems. Distribution of electricity is the local transport throughout the streets, back alleys, and rural highways across the world. Chapters 4 and 5 will cover how utilities use GIS in distribution in detail. Street lighting is often included as part of the distribution system, but in many cases street lights are owned and operated by the local municipality, which can create challenges for the electric distribution provider. Utilities rarely model high voltage to medium voltage (HV/MV) substations (i.e., those that tap off the transmission system in their GIS). They should be.

Customer Care Needs GIS

The next link in the value chain is the retail sale to the end use customer. This will be covered in Chapter 7. These will be the customer care processes. Every customer is located at a different place. GIS can play a vital role in managing the new smart meter rollouts, by supporting customer analytics and helping utilities understand their customers.

GIS Is Engaged in the Complete Utility Value Chain

Figure 1.9 shows the typical electric utility value chain. The energy supply organization manufactures the power, the transmission organizations delivers it in bulk, the distribution organization taps the power from the transmission system for local delivery, and the customer care organization sells them the power and collects the money. Chapter 6 details emergency management and is covered separately because it often involves the integration of a number of different groups within the company. In addition, there are several other support organizations, such as supply chain, finance and accounting, and corporate services all included as part of shared services in Chapter 8.

GIS Is Not Just About Mapping

Electric utilities have used GIS for quite some time, yet the vast majority of its use has been for modeling the electric distribution system and mostly limited to the medium-voltage feeders and some of the lower-voltage equipment. While this has proved to be successful for providing better and clearer maps, its common application still falls far short of its full potential to positively impact the big problems of the utility—the things that really matter. Given the challenges utilities face, they need to leverage every tool they can muster.

What About the Data?

Most electric companies have a data problem. They have too much data, and the data is in different forms. They have authoritative data stored in the customer relationship management system (CRM), financial, and enterprise asset management

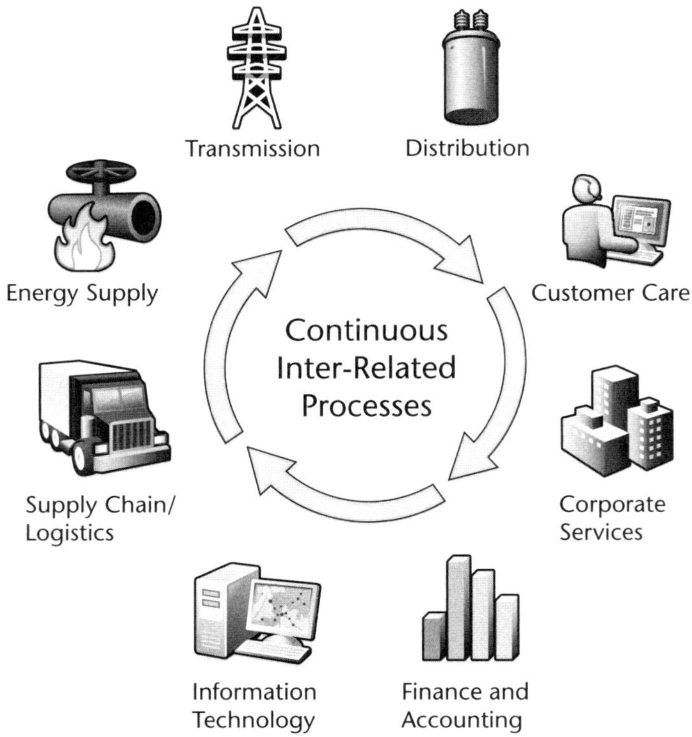

Figure 1.9 Utility value chain.

(EAM) systems. They have predictive information, like flood data or lightning strike data available but not in an easy to use form. For those that installed smart grid and advanced metering infrastructure (AMI), they will have an order of magnitude increase in consumption and power quality data. They also are getting data from social networking sites. Finally, they continue to capture normal real-time data from their control and monitoring systems such as SCADA and DMS. They also have data locked up in their workers' minds with no easy way to capture the information. The business impact is that they may be making decisions based on an incomplete understanding of the situation at hand.

Since utilities have so much data, it's hard for them to consistently perform an analysis. The business impact is that they make decisions based on worker experience. As knowledge workers retire, that insight will be lost. Given the volume of data and the lack of analytics to deal with the data, it becomes difficult to communicate and collaborate with the work force.

GIS Helps Solve the Big Problems

The GIS capabilities of data management, spatial analysis, and awareness help lessen the problems facing utilities if they would take advantage of those capabilities. Since the vast majority of that disparate data is spatial, the GIS can sift and sort through it, reconcile the various projections and accuracy levels, and provide a common view of the data sources. Then, once the data is organized and coordinated, GIS can provide a whole host of spatial analytics to lend some clarity around what exactly is going on. Figure 1.10 shows a common way that GIS takes many data layers and analyzes them to create a hot spot map. This map distills all the factors into a single display. The hot spot clearly stands out as a place that may be the most vulnerable to a storm or the place where customers are unhappy.

Finally, GIS is designed to display the results of the data fusion or analysis on a digital map. Data fusion simply means that data from different sources are combined together on a common format (like in a map) for the purposes of seeing the relationship of many factors to each other. That map can be on a desktop, the web, or a tablet or smart phone. The idea is that once insight is generated (e.g., from a spatial analysis model or from a red line on a smart phone), information is disseminated throughout the organization. A convenient delivery mechanism is the cloud. Once information is stored, any authorized user can access it.

Need to Complete the Data

So utilities have grown from using hand-drawn maps (although there are still a surprising number of hand-drawn maps still in use at utilities around the world). They converted their hand-drawn maps to a digital format, but primarily for ease of drafting, not as a strategic information resource. Utilities are beginning to recognize the strategic importance of the GIS. Now that they have the right tool in place, the challenge will be to fix the data. Many utilities still have not modeled their structural data in the GIS (the conduits, manholes, and vaults). Many have not migrated their land information to a GPS-compatible base map. This also means that their facilities in their mapping system are not in the correct position from a GPS perspective. Often, phase designations are missing.

Many utilities take too long to get the data from field into their GIS. In their haste to take advantages of improved productivity from their mapping departments, they trimmed their staffs before they tackled the backlog of information

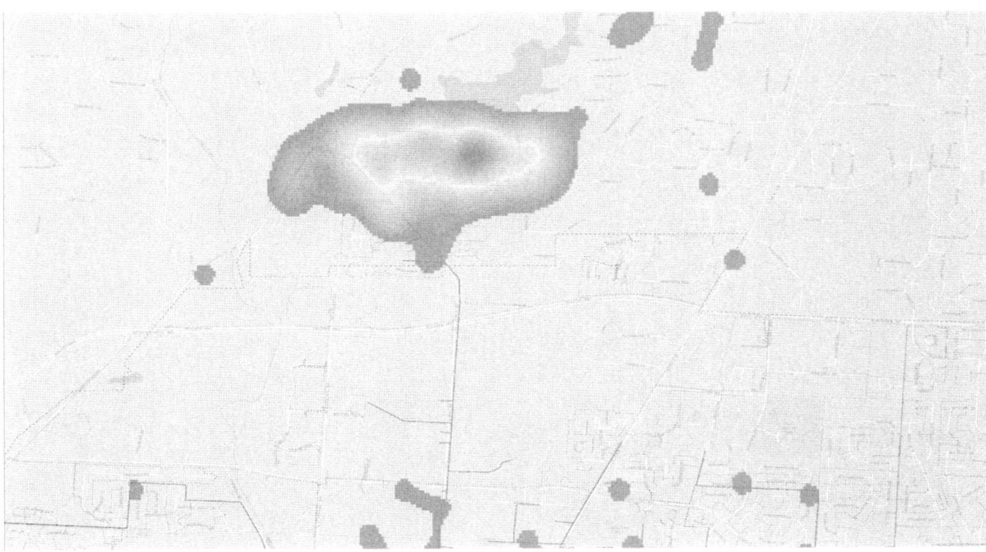

Figure 1.10 GIS hot spot map (Source: Esri et al).

that had not made it into their GIS. Using mobile GIS in the field can significantly shorten the cycle time from when something has been installed to when the data about that change in the field gets posted into the GIS. Finally, many utilities have kept their GIS in the back room of the mapping department and have not embraced the notion of the GIS as one of just a few critical information systems.

GIS architectures are now fully embracing the new technologies of the cloud and the mobile devices. Thus, utilities are opening up the GIS to the electric utility enterprise, not just the few in the drafting department.

CHAPTER 2

Location Matters for Energy Supply

When people think about an electric company, they often think about the actual product they are consuming, namely, electricity, and not so much about how it gets to them. Here, we will deal electrical energy supply and how location plays a critical role in the development and management of the sources that produce electricity. Since location matters, GIS plays an important role in the management of the production and development of electricity sources. Unlike much of its history, the utility business today is actively looking for new ways to produce electricity. Two factors, climate change, due to the carbon dioxide emissions, and security and availability of fuel sources drive the need to diversify supply.

Given that all forms of fossil fuel, from petroleum to coal to natural gas are nonrenewable (nuclear is nonrenewable as well), communities and regulators are demanding that utilities produce or procure electricity made from renewable resources. While location matters in the management and development of conventional sources of electricity production, location is perhaps even more prominent when dealing with renewable resources, given their variability and the strong relationship of these sources to the environment.

What Is in This Chapter?

This chapter discusses the changing nature of the generation business from one dominated by vertically integrated utilities to competitive entities and how this has changed the business. Further, this leads into why the use of geospatial technology can help them cope with the new challenges that come from a shift in the business model.

The next several sections deal with how GIS can positively impact performance of the energy supply ecosystem: central plants to renewable energy sources. The chapter continues with how GIS deals with the actual development of the energy sources. A discussion on how GIS impacts the energy supply organizations follows. Finally, the chapter introduces the notion of an information model.

A New Era for Electrical Energy Supply

The business model of the utility itself has changed dramatically over the last decade or so. There has been three major epochs in how electric companies operate. The first era is the free-for-all period. During the formative years, electric companies sprang up all over the world. In fact, there were battles over the technical nature of the business. The most well-known battle was between the advocates of direct current (dc), spearheaded by Thomas Edison, and those who believed in the benefits of alternating current (ac), led by George Westinghouse and his inventor scientist, Nicola Tesla. A number of players entered the market, installing poles and wires in a frenzy to capture the market. In the free-for-all era, the early electric systems were simple. The early electric companies operated generators or dynamos with wires connecting directly to the end users. The result of all of this development was that competing systems cluttered city streets. As electric service became more developed and people relied more and more on it, governments recognized that having duplicate infrastructure was getting intrusive.

Moving from the Free-for-All Era

The notion of granting a single supplier of electricity an exclusive franchise territory appealed to governments around the world. Two models emerged: private and public. Many municipalities opted to simply buy out the electric producers and own and operate the electric system within their borders. However, the notion of government-owned businesses runs contrary to the principles of capitalism.

Other communities granted franchises to private electric companies. The problem, however, was that these private companies would have to be monopolies. Capitalistic societies do not like monopolies either. So communities drew upon past business models set up for grain elevators, ports, and some railroads called the utility model. To protect the public from arbitrary price gouging, communities established regulating bodies that would require the utility company to submit prices for approval of the community. Thus, a privately owned company struck a regulatory deal with the community (in the U.S., the community was the state, since at that time electric companies rarely did business across state lines) that granted monopoly status (thus assuring no competition) in exchange for a fair and reasonable rate of return on the utility company's investment. The final arbiter of any price increase would be the community. So this did a few things:

- It created privately held regulated monopolies.
- It established community-based regulating bodies. In the U.S., for example, these bodies are often called public utility commissions. These bodies regulated rates charged by these private companies.
- These changes ushered in the era of the regulated privately held vertically integrated utility.

Throughout the world, vertically integrated electric utilities were either investor owned (privately held) utilities or publicly owned (municipal, federal, tribal, special district owned). Electric cooperatives are a variation on publicly owned utilities. Instead of the government, which represents consumers, owning the utility, customers own the utility. Communities tend to think of electric cooperatives as publicly owned utilities.

Competition for Energy Supply Emerges

The long-standing assumption of the vertically integrated regulated utility era was that duplication of facilities for the purpose of competition was not in the public good. However, over time, with the development of the transmission grid, regulators began to question this. Why, they argued, would duplicate generators not be in the public good? Each generator could plug into the transmission grid, creating a market, just like a food or flower market. Those generators that were efficient would win; the inefficient plants would lose. A vibrant competitive market would spur innovation, and the beneficiaries would be the consumers. Regulators across the globe restructured the electric market to unbundle vertically integrated utilities into the component parts, generation, transmission, distribution, and retail.

The Era of Competition

This ushered in the current era of electric utility liberalization, which is characterized by these changes:

- Generation shall be competitive.
- Markets for generation shall be set up.
- Transmission shall continue to operate as a regulated monopoly, but operators shall be required to allow all generation operators open access to the transmission grid, charging tariffs that are the same across all market participants.
- Distribution companies shall operate in a regulated monopoly as in the past, but may be excluded from offering metering and billing services.
- Competitive retail companies may be set up to offer consumers electricity service (not transmission or distribution services) at whatever price they want.

This era is still under development. In the U.S., not all states implemented industry unbundling (often incorrectly called deregulation). So where this regulatory process is in effect, the role of energy supply (power generation) is now different. In the old vertically integrated utility era, as long as utilities built power plants reasonably, kept good records, and operated the plants well enough, utilities by virtue of the regulatory deal would earn a return on their investment. In some cases, they would earn a return whether they actually operated the plant or not. In the new world of unbundling, only those plant operators of plants that actually produced electricity would make money. This is a bit oversimplified, since the electricity market is quite complex, but the notion is clear. Operators of power plants in this new era are subject to the rules of competition, just like oil refineries or steel mills.

To deal with the generation as a competitive market, governments established electricity markets. Figure 2.1 illustrates a simplified diagram of how electricity

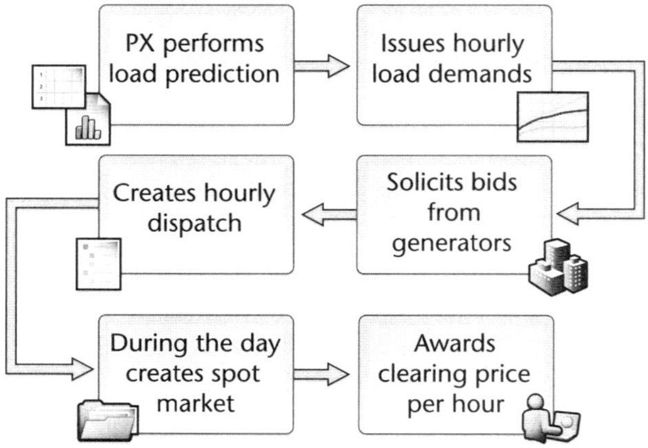

Figure 2.1 Idealized electricity market operations.

markets works. So generators will be looking for every conceivable advantage to keep costs low and mishaps few. (Note that PX stands for power exchange, and is the entity that manages the power market.) Operators can predict or even avoid many of the mishaps by a careful assessment of what is exactly going on around them and at the plant itself. Most of what happens at the plant has something to do with location.

Shifting of Risk

In effect, the new regulatory framework has shifted the investment risk from the consumers or rate payers to the owners of the plants. At the same time, if the market were to function completely unregulated, the consumers now are at risk of price instability. Given these new risks, all aspects of the business needs to meticulously scrutinized. With a misstep anywhere along the way, the investors could face serious financial consequences. What if the plant shut down for long periods of time? What if they spew greater than allowed emissions? What if something compromised the cooling water systems resulting in environmental damage? What if there is a security breach? What if the plant runs at less than optimal efficiency? What if a natural disaster shuts the plant down? What if the wind doesn't blow or sun doesn't shine according to predictions?

Under the old regulatory scheme, these things would be bad for sure. Under the new regulatory framework, these events could be disastrous. On the other hand, if an operator owns a plant that has stunning efficiency and high availability, it will generate a high profit for its owners. If operators had to raise prices significantly, say due to fuel shortages, natural disasters, climate emergencies, or the inability to meet demand, communities may push toward returning to the old regulated model, which would create new uncertainties in the marketplace. While profit will be a strong motive, operators will be reluctant to drive up prices as well. Given this new regulatory framework and the drive toward renewable energy sources, utilities (or power plant operators) need all the help they get to gain a competitive advantage.

The use of GIS is one of the tools that can help, since GIS can help utilities articulate risk using spatial analysis.

The Energy Supply Market

While most of the world's electric production comes from fossil fuels, this mix will certainly change over the next several decades.

The Market Will Only Get More Complicated

Aside from the shift in risk, another shift will involve location. Today, the vast majority of electric generation plants are large. The model consists of thousands of very large centralized power plants, each plant having a huge geographic footprint. Some are as large as small cities. These large plants then feed into a huge interconnected transmission system. Tapped off the transmission system are the various load centers. This is the present model.

In the future, there will be more decentralization of generation sources. These sources will be mix of renewable energy sources to be sure but also of smaller conventional generators, powered by fossil fuels. This will result in the emergence of smaller independent grids, able to sustain themselves during a power failure. Utility people use the term microgrid for a self-sustaining electric supply and distribution system that can continue to operate while disconnected from the large utility grid. The microgrid could be as large as a city or as small as a single building or a college campus.

Energy Supply Gets Even More Complicated

These factors mean that the energy supply market will be much more complicated in this new era. Instead of market operators having to keep track of thousands of very large plants, they will need to monitor perhaps millions of small plants. In this new world of microgrids and distributed generation, location of those point sources will be paramount. Figure 2.2 illustrates how GIS can display the sources of small distributed generators. While every power plant produces the exact same product, electricity, each type of plant has its own uniqueness, advantages, and problems.

The Energy Supply Organization

Whether a utility actually owns and operates the actual production of electricity, they still are responsible for procuring the power to deliver. Figure 2.3 shows the various processes involved in a typical energy supply organization of a utility. In addition to the actual production, energy supply organizations buy and sell power, issue power contracts, look at their financial risks, and hedge their purchases. Even though many utilities have divested their generating assets, most electric companies still have to manage their energy supply portfolio.

Energy Contracts and Risk Management

Utilities will need to understand their sources of supply and the potential risk of losing that supply. Loss of supply is primarily due to outside factors, such as weather events, natural disasters, vandalism, or terrorist attacks. So if a utility has entered into a long-term supply contract with an independent power producer, they will

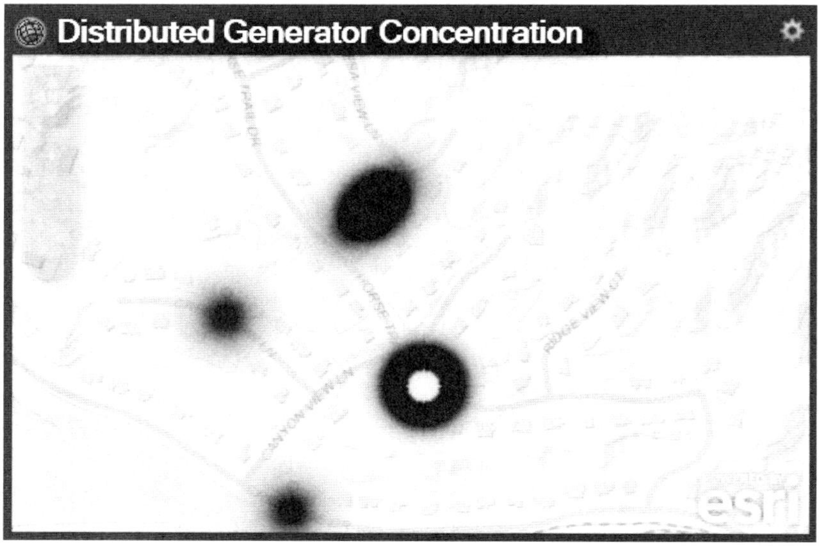

Figure 2.2 GIS maps shows concentration of distributed generation. (Source: Esri et al.)

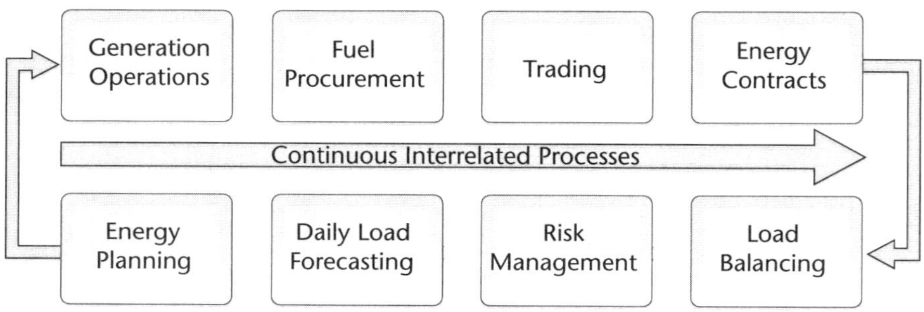

Figure 2.3 Major processes of an energy supply organization.

need to understand the chances on a given day or hour what the risk of losing that energy is and make adjustments for those risks. In fact, in the determination of the contract itself, the utility must assess the risk profile of that plant to better understand how much to pay for the power contract. They will need to build the risk profile into the terms and conditions of the power contract.

There are a number of risk factors to be looked at that are not spatial, like age, historical availability, management controls, safety records, compliance records, and emission history. However, risks are also associated with its location. When making long-term power contracts, utilities can better determine the relative risk of an existing or potential power source by combining the nonspatial risks with those that have a spatial component to determine the relative risk of one plant verses another. Spatial risk factors include things like proximity to earthquake faults, tsunamis, rising tides, and landslides, and the proximity to a whole host of possible disasters. There are factors like crime, proximity to known terrorist cells, social uprisings, floods, or any other factor that could impact the availability of the power.

Figure 2.4 shows how GIS can take nonspatial and spatial data together to develop a map called a hot spot map that illustrates where power plants are most vulnerable or have the highest risk of something bad happening to them.

Like other parts of the utility value chain, those that own and operate generation plants have to manage the performance of the plants, more so under the new rules for competitive generation than ever.

Managing Power Generation Performance

Each type of power plant has its own unique characteristics that impact on its performance and cost effectiveness. At the end of the day, a power plant's performance depends on its fixed costs, like the amortized cost to build the plant in the first place; its operating costs, like labor and materials to run the plant, which may be more or less fixed; and its incremental or variable costs, which typically are the fuel costs. Other important considerations are its up time or availability—the volume of power produced annually, how fast a plant can start up when called upon, and the environmental impact. As the world moves closer to a carbon cap and trading system or a carbon tax situation, the cost of emissions will dominate the business model for a particular power plant. The better the plant operator manages its data, analytics, and data dissemination, the better the operator can make critical investments and operating decisions. GIS can be a part of that.

What follows is how GIS can improve the various operations of different types of generation.

Figure 2.4 GIS map showing risk profile as input to energy contracts. (Source: Esri at al.)

Hydroelectric Generation

Hydroelectric generation systems are examples of power plants that are very expensive to build, have modest operating costs, and no fuel costs. A pumped storage hydro plant is a variation on the hydroelectric plant. That plant uses hydro power from an upstream water source to generate power during periods of high demands, then during periods of low demands, normally at night, reverses the process and pumps water back behind the dam. The consequences of a failure or misstep in the operation of a hydro plant or a pumped storage plant can be catastrophic. Should a dam break or overspill, massive flooding could occur, threatening whole communities while destroying wildlife and raising long-term havoc with the environment. Figure 2.5 illustrates how GIS is used to model the impact of a dam break. That is why hydro plant operators carefully model their plants in GIS to better understand the potential flood impacts. They use GIS to model upstream resources and downstream impacts.

While governments and communities consider hydroelectric plants renewable, the operators have to be concerned about methane emissions, which is a greenhouse gas. Since hydroelectric plants dam water and in many instances flood areas of former vegetation, the vegetation will decompose under the flooded areas and emit methane gas. So a hydroelectric plant operator needs to monitor the emissions of methane and know exactly what type of vegetation formerly existed in the areas now holding the water. GIS models the flooded area and the former vegetation as well as illustrates the emission of methane gases.

Since the only fuel used to produce electricity is stored water, it is essential for the hydroelectric operator to model exactly the "fuel source" or the storage of the water, based on infiltration, rainfall, soil types, evaporation, and a host of other factors to make sure that the operator can predict the production of the plant. A hydroelectric plant operator who participates in the power exchange market must be able to bid into the market like every other participant. Figure 2.6 illustrates how GIS works in concert with a production-planning model to determine the

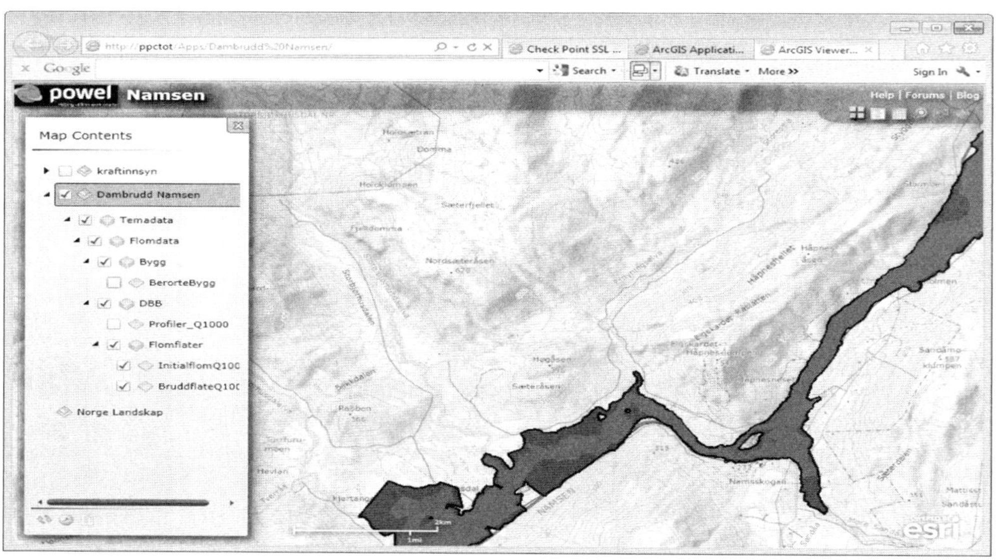

Figure 2.5 GIS map determines the impact of a dam break. (Source: Powel Corp. and Esri.)

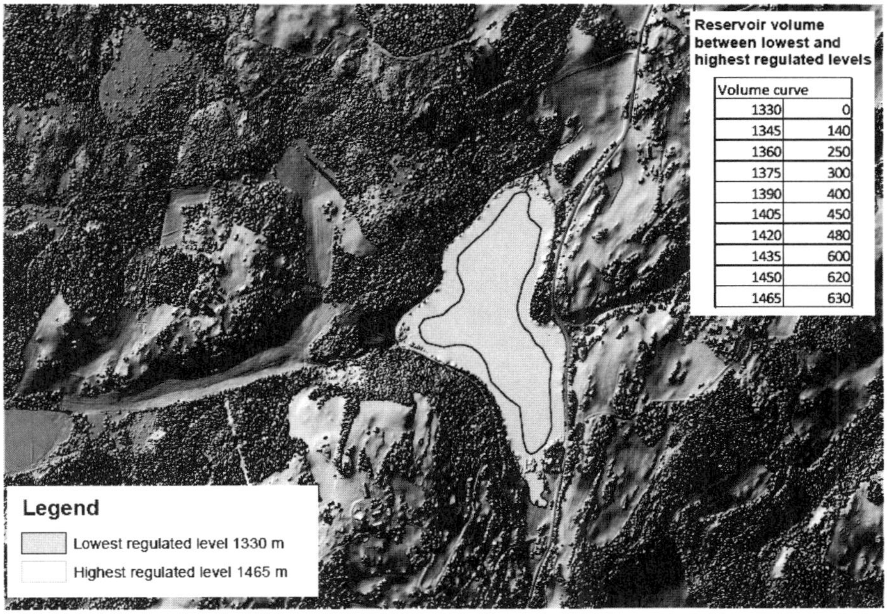

Figure 2.6 GIS map showing volume levels as part of a production-planning model. (Source: Powel Corporation and Esri.)

storage volume for water. The GIS determines the storage volume by merging a digital elevation model with imagery of the terrain along with polygon of the water extent, plus real-time readings of the water level.

Unlike, a fossil plant operator who knows how much coal is stored in a coal pile or natural gas is stored in a tank, it is more complicated to exactly determine production based on so many environmental considerations. Operators need to create a detailed model of all the land features in the GIS and adjust the model for shifting conditions. Hydroelectric plant operators have to have ready access to weather data, so they can accurately predict their "fuel."

Hydroelectric plant operators, like operators of other kinds of power plants, must assess the environmental impact continuously. Since most plants have to undergo relicensing on a periodic basis, they must be diligent in capturing the impact of the plant on the surrounding land and communities. The GIS provides the most complete picture of the plant and its impacts.

Nuclear Plant Performance

Unlike hydroelectric plants, nuclear performance is very predictable (provided nothing goes wrong). Operators can add new nuclear fuel only during a plant shutdown. Nuclear plant operators call this down time a refueling outage. Under ideal conditions, the plant runs continuously at full production until the fuel runs out. Since the variable cost of production for a nuclear plant is very low, the ideal situation for the power exchange and for the operators is to dispatch (select to run) every nuclear plant available. The most critical issue (aside from avoiding an accident) for a nuclear plant operator is to absolutely minimize the refueling outage time. It goes without saying that an unscheduled outage is devastating to operating revenues. Every hour the plant is not producing electricity, the operator loses vast sums of money.

So from a performance perspective, the nuclear plant operator needs to keep track of every possible situation that could shut the plant down or force the operator to curtail production. The use of GIS for performance then is more about risk management than direct production planning as in the case of the hydroelectric plant. Risk management would include modeling the impact of any event that could interrupt operations and subsequent restart or delay the refueling outage. Operators need to model every conceivable risk that could occur during a refueling outage. Even a simple wind storm or heavy rain could impact the schedule of a refueling outage. Logistics are critical. Operators must carefully optimize every movement of every piece of material and personnel. GIS's ability to model linear networks and optimize routing can be essential in the planning of the refueling outage. A one-day delay in the refueling outage of a typical nuclear plant could cost the operator millions of dollars.

GIS Helps Nuclear Organizations to Plan for Emergencies

Nuclear plant operators must keep ready access to the land features of the plant and surrounding communities should an abnormal event occur. The use of GIS for direct nuclear operations includes:

- Modeling evacuation zone and routes;
- Determination of severe weather on all equipment;
- Earthquake and tsunami threats;
- Wild fire threats;
- Flooding;
- Loss of electric transmission and backup power to plant;
- Cooling water interruption;
- Contamination of the water supply—modeling of ground water and proximity to rivers and streams.

The nuclear plant operator will be most concerned about modeling adverse impacts and crafting contingency plans should something bad happen. Figure 2.7 shows a map of where nuclear plants are located. This helps communities understand the impact of a nuclear evacuation should that be necessary. Operators can base their contingency plans are on GIS since location of events, evacuation routes, security breaches, and a whole host of other abnormal events involve location. Even during a routine refueling outage, operators may want to model the plant facilities in GIS to optimize the movement of people, materials, and machines.

GIS's most effective tool for the determination of risk is called a spatial analytic model. This model will be illustrated later in this chapter for site selection. The model consists of three elements: spatial and nonspatial data sets as inputs; geoprocessing tools such as overlay, union, intersection, and buffer that operate on the inputs; and spatial representations of the results of the operations. Evacuation analysis involves modeling a number of scenarios, such as wind direction, road

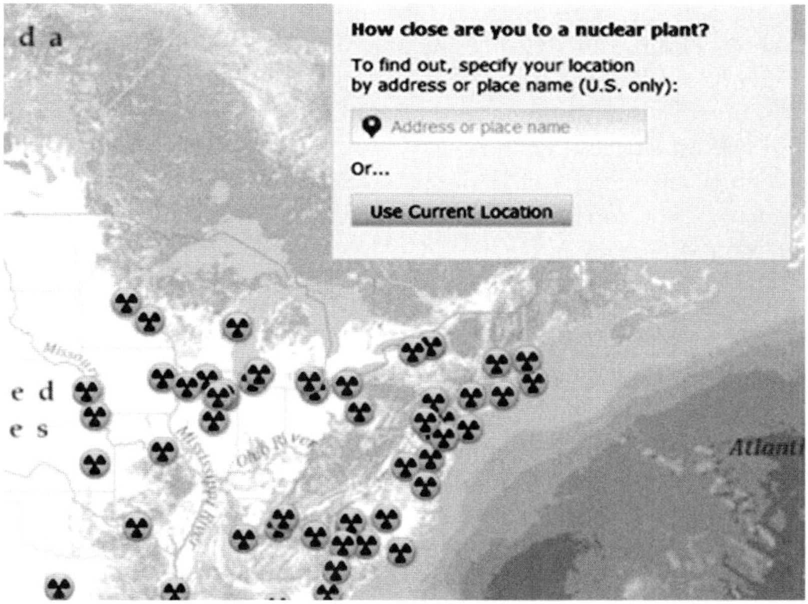

Figure 2.7 GIS map showing nuclear plant locations. (Source: Esri et al.)

capacity, weather, terrain, and a host of other variables depending on the reason for the evacuation. Evacuation planners must be able to model a nuclear accident under all kinds of disaster scenarios. The recent earthquake and tsunami in Japan illustrated how many different factors can complicate evacuation scenarios.

Fossil Plant Performance

The design of the plant and its operation determines fossil plant performance. Just like nuclear plant operators, fossil plant operators must minimize plant down time. Unlike a nuclear plant, down times are not dependent on the fuel running out. Rather, operators schedule down times to perform major maintenance activities. Regardless of why they shut the plant down, operators will need to optimize the logistics in moving materials, staging activities, and any unscheduled activity to absolutely minimize the down time so that the plant can resume operations. Gas turbines have normal down times during off-peak times. Even then, significant cost penalties can accrue if the plants are unavailable when called upon, particularly during periods of peak clearing prices. Logistics around how well and fast maintenance is performed during a routine outage of the plant will impact the overall profitability of the plant. Also, as in the operation of hydroelectric and nuclear plants, planning for extreme weather events, earthquakes, loss of transmission, terrorist attacks, and sabotage need to be planned for, and most of those plans require a solid understanding of location.

Managing the Risk

Unlike nuclear and hydroelectric plants, the security of fuel supply is a critical issue. Fuel sources usually depend on third parties, like natural gas pipeline operators or railroads that haul the coal. Interruptions in the fuel supply will shut the plants down. So utilities need to be aware of the risks of those interruptions and model them in the GIS.

Emissions monitoring is essential for continued operations of the plant. If emission control devices fail and emissions violate regulations, regulators will fine plant operators and ask about the impact of the emission failure. Fossil plant operators need to continuously model current weather patterns and be able to model the emission violations in the GIS. Coal plant operators have to be particularly concerned about possible ground water contamination of the coal piles. So monitoring the proximity of water resources near the coal storage facilities is essential. To do this, planners capture the water resources as a GIS data set and create a spatial analysis model that determines the proximity of water resources to possible leaching areas from the coal sources.

Some coal plant operators own and operate the coal mines that supply coal to the plant. Utilities have built power plants at the mouth of the coal mine itself. So they may need to model the logistics of the coal supply and optimize its transportation.

Rail operations are critical to the fuel supply of coal plants. GIS is used extensively to model railroad facilities and operations. Coal plant operations can work closely with railroad companies to access information from the railroad GIS and include the data sets as part of their spatial analysis and routing models for coal supply. There are a number of resources available that deal exclusively with GIS and railroad operations.

Performance of Wind Power

Like hydroelectric plants, nuclear plants, and many fossil plants, wind plants are expensive to build, and do require a fair amount of maintenance, but wind farms have no fuel costs and of course emit no greenhouse gases. Unlike nuclear, large base-loaded fossil plants and to large extent hydroelectric plants, the power generated from wind plants is variable. That means that production does not follow the demand. While the power production is variable, it is not entirely unpredictable. Wind power plant operators can use sophisticated wind forecasting models to help them in the daily load forecasts. The topology of the land, weather models, and historic wind patterns determine much of that analysis.

These GIS prediction models use imagery and digital elevation data sets, along with accurate representation of the location of each of the wind turbines. Offshore wind farms must have a solid understanding of the topology of the ocean floor, wind patterns, and the constant shifting of the ocean floor that occurs. Figure 2.8 shows the GIS map of the ocean floor.

Like every other type of power plant, operators need to maximize the investment in the wind plant. That means the operators must optimize the maintenance of the plants. The ideal situation would be to perform maintenance of each turbine during a time when there is low demand and when the GIS predicts the fuel source, the wind, to be the lowest. Since a wind power plant consists of many separate wind turbines, it is necessary for operators to have each turbine located exactly so they can perform the necessary modeling of wind prediction. Operators can use

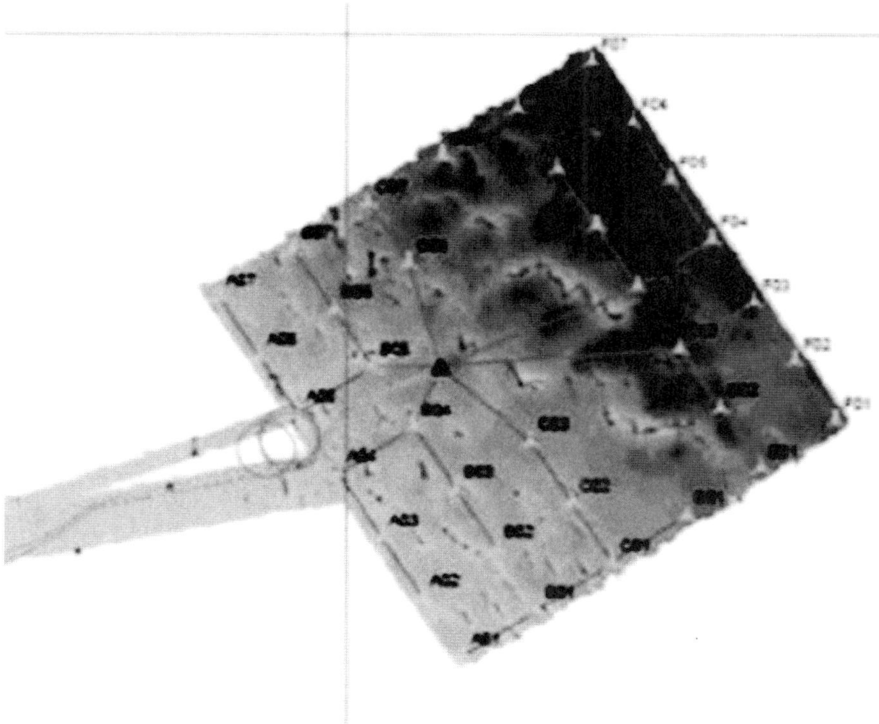

Figure 2.8 GIS map of the ocean floor for offshore wind farm modeling. (Source: Esri.)

GIS to perform optimal routing of field crews for inspection and maintenance to limit the downtime of each of the turbines.

With the purpose of maximizing availability, offshore wind farm operators must account for and model responses for unscheduled events due to extreme weather, sabotage, and errant ships with their anchors digging up cables entering the offshore wind farm areas. Here the GIS can model the perimeter of the offshore turbine locations and alert operators if a vessel enters the space of the wind power plant. Geofencing technology can be used to alert operators that a ship has entered within a range of a wind turbine platform. Geofencing uses sensors along with GIS that is like a virtual trip wire. Once a vessel enters into within a polygon around the wind turbine, the operators know about it immediately and can take action.

Wind Still Has Environmental Issues

Wind plants do not emit greenhouse gas, but they still pose an environmental impact. GIS can model them in order to stay in compliance and assure smooth performance. There are these factors:

- *Noise*—noise level profiles can be overlaid on base maps of the region to determine the exact decibel levels near communities. This will be important for compliance and to avoid community disputes.

- *Birds*—a common problem with wind farms is the interruption of bird migration patterns and the killing of birds. While bird migration patterns should have been dealt with during the siting process, many wind power plants continue to be plagued with bird issues.
- *Strobing*—the rotation of the blades produces a sort of strobe light effect that can be very annoying for those that live near the turbine. Stobing can be modeled in GIS by examining the sun rising and setting and then creating an overlay where and when strobing is most severe.
- *General aesthetics*—this issue is usually addressed during the siting phase. However, there still could be ongoing community concerns about the obstruction of views that a wind power plant operator must address.

These issues are all location based and are part of a GIS model. Complaints about excessive noise, bird killings, and strobing in the worst of situations can shut down the operation of turbines and can have a devastating impact on the performance of the plant. Having these issues accurately modeled helps operators work with the communities to mitigate the risks.

GIS Data Management Helps Wind Farms

When dealing with complex issues of siting a wind farm or any large facility, one of the most difficult processes is the gathering of up-to-date and complete data. In the case of a wind farm, developers must gather data from so many different sources. The challenge is to make sure that the sources are current. Most developers today still rely on the old process of gathering printouts of charts and maps, and then attempting to piece together all of the factors. Of course, this process takes time. While a critical piece of data or a map might have been current when the developer got a copy of the map, there is no way of knowing that the data on the map could have been updated. However, using a GIS spatial analysis model that accesses critical map data, like bird migration data using a web service, the developer will know that each time the analysis is run, the data used in the analysis will be the most current digital map data available.

Today agencies, governments, and think tanks publish their data over the web and store it in a private or public cloud. So access is searchable from the GIS and quickly accessible to the developers.

Solar Electric Power Generation

Like wind and hydroelectric power plants, solar plants are expensive to build, require some level of maintenance, and have no fuel costs. From a market perspective, these forms of generation have the lowest incremental costs. The vast majority of solar plants are small photovoltaic installations on rooftops and open fields. However operators are building and operating a number of large-scale central solar plants. Passive solar plants use mirrors to heat a fluid. The plant then pumps the fluid deep into the ground. The fluid is stored and used later.

Aside from maintenance to keep the PV panels clean and the electronics running, there are few environmental factors inhibiting the use of solar generation. For very large PV installations, it is important to keep track of the production of each of the panels, so that should a panel fail, the operator can easily locate and replace the failed panel. GIS models the exact location of each of the panels and can provide optimal routing for cleaning, replacement, and maintenance crews.

Given the distributed nature of some large-scale solar installations, utilities will need to know where they are located. There is an emerging business model for small-scale solar panels. Developers or even utilities provide solar systems to homeowners and small businesses without cost. They continue to own the solar panels, but provide a discount to the homeowner or small business for the right to install the solar panel on their property. As this process becomes increasingly commonplace, utilities will need to know information about these assets and exactly what these utility-owned panels are generating. The locations and attributes of these panels will be captured in the GIS.

Geothermal Power Plants

Geothermal plants, like hydroelectric, wind, and solar, are expensive to build, require a fair amount of maintenance, and have no fuel costs. Geothermal plants rely on the heat well below the Earth's surface for its energy source. This requires the pumping of a treated fluid deep into the ground and brought to the Earth's surface to generate electricity. The cost of a geothermal plant to build is directly in relation to how close to the surface of the Earth the geothermal resources lie. Most large-scale geothermal plants are located near areas of active volcanos and hot springs.

Once built, the plant operates like any other power plant and is subject to the same threats to production as any other plant: extreme weather, earthquakes, sabotage, and so on. Since the plant requires that a heat-conducting fluid be pumped deep into the Earth, operators must be concerned about any aquifer contamination should a pipe fail. So operators need GIS to model the sensitive underground water resources.

The recurring theme for most of the generation resources is that utilities need to understand the location of the plants along with risks and hazards near the plants. Utilities need to have spatial analysis models that calculate various risks should something happen, like in a geothermal plant if the conducting fluid pipe should rupture. The spatial analysis model determines how close the rupture is from a sensitive aquifer or other environmentally sensitive area and is able to determine the impact of the damage.

Distributed Generation

With the growing investment in advanced metering infrastructure (AMI), utilities will shift their pricing from average monthly pricing to dynamic pricing. Utilities will strive to match the production cost (based on the clearing price for that time interval) to the price they charge consumers. The implications are that the cost of electricity at peak usage times to customers could rise substantially in this model. Today, utilities charge commercial and industrial customers for both an energy and power demand. They often base demand on the peak use of power during a particular billing cycle (normally one month). In the new model, the price of electricity could vary not just at the peak, but during periods of high production cost, (e.g., during periods of draught or low wind production), which may or may not occur during a conventional peak time. The result will be that more and more consumers will install on site generation in an attempt to manage the dynamic pricing variance.

Utilities will need a very precise model should distributed generation proliferate. They may need to monitor the production of this generation and certainly need to understand the location. They will also need to include the distributed generation resource models in their GIS. Many utilities have not focused very aggressively on capturing the data about their low voltage network (operating below 1000V). However with more and more distributed generation, utilities may find that the low-voltage networks may not be able to handle this additional stress. Widespread distributed generation (including wind and solar) plus an increasing number of electric charging stations could render the low-voltage network very vulnerable. There will be more on this in Chapter 4.

Energy Planning

With all these types of generating sources available, utilities and system operators will need to know in detail where all these sources are located and what the characteristics of each type of source are. With a carbon tax or carbon cap and trade program in place, operators will need to understand the financial considerations as well as the load balancing considerations. This will become increasingly complex as the percentage of variable sources of renewable generation becomes higher. One of the components of this complex analysis will be the matching of fairly detailed load forecasting models to the generation mix.

Today, the vast majority of generation sources are centralized, pumping power into the transmission grid. The vast majority of the world's power requirements were met almost entirely from centralized generation plants. Each generator automatically adjusts for the increase or decrease in the demand. That model will shift to more and more decentralized generators, many variable, with automated demand response systems in place. To further complicate the situation, storage facilities will become more prominent, dispersed in communities, creating the possibility of smaller independent grids that sometimes are connected to the main grid and sometimes not.

So it will become essential for utilities and system operators to know the location of all the small generators and storage facilities. They will need to be modeled in the GIS.

Managing the Infrastructure

The demand for electricity will continue to rise in the years to come, even though the industry has seen demands stagnate over the last several years due to the great recession. As the economy rebounds, the use of the electricity will increase. In addition, there will be shift away from petroleum to other sources of energy for transportation. A simple calculation reveals that if everyone in the world were to convert from petroleum-fueled vehicles to electric vehicles (of some kind), the amount of raw electrical energy required would be about half of the entire worldwide electricity production. Clearly that's not going to happen anytime soon. Yet suppliers of electricity need to be aware that as the transportation industry looks more to electricity for fuel, the demand for electricity could rise substantially. A significant increase in the price of petroleum over a long period of time will also accelerate the shift to electric forms of transportation. That shift would include freight rail perhaps as well, which for the most part is completely dependent on petroleum.

The problem for society will be producing enough electricity to supply the needs of the future. The drive toward renewable energy and the reduction in greenhouse gases will move operators of generators to continue to build all forms of generation. They will replace those existing generators that will fail to meet carbon emission standards in the future.

What's Involved in Building Electric Generation?

Two phases result in the building of a new electric energy source:

- *Exploration*—this is the process in which the optimal source of energy is discovered;
- *Development*—once the source of energy for the production of electricity is determined and validated, the development of a facility (such as building a wind farm, for example) is undertaken.

Exploration

Exploration involves finding the right place to develop a source of electrical energy. In the case of geothermal energy, for example, the exploration process involves a search for those areas where there exists significant seismic resources (i.e., areas where significant heat from the below the Earth's surface is relatively close to the surface). Another example is the exploration of solar resources. Figure 2.9 shows a GIS map of solar density, which is really a map of various solar resources. During the exploration process, the GIS is used to determine where the resources are and the resources in relation to other factors that could determine if the resources are worthy of development. The GIS then uses spatial analysis to accumulate other factors, such as weather conditions, environmental situations, the topology, the proximity to electric transmission, the climate, land availability, and even social political situations.

The exploration steps are to first determine or discover the resources using GIS and then analyze the suitability of developing those resources. Once developers discover that a resource is suitable, a detailed siting analysis follows. It's critical that the information used to discover the resources and the suitability is maintained and accessible so that later, during a challenge, the data used in the analysis is readily available during the development of the plant itself.

A simplified example of site selection spatial analysis would be the answer to the question, where is the best place to build a solar generator assuming we only have two criteria?

- The highest solar density;
- Nearest to an electric line.

Figure 2.9 GIS map showing solar densities. (Source: Esri.)

In this simple example, the planners have a spreadsheet of solar readings spaced at different intervals within the study area. The spreadsheet includes solar intensity as a column and GPS location as another two columns. The first step in the spatial analysis is to convert the solar readings to a solar density map. The spatial analysis tool selects a cell size for the analysis, roughly selecting a square cell where the center of the cell is the location of the reading. Obviously the more readings and the nearer the readings are to one another, the better the density map. This is an example of a density analysis tool. The tool converts discrete points into cells or pixels that are uniform across their area. In effect, the analysis converts a series of points to a raster image of a series of pixels.

The next step is to normalize the density map from the solar readings themselves into a ranking, such as 0 to 100. The reason for doing this is to be able to compare the solar density factor to other factors, like land suitability. The normalization tool assigns each cell or pixel a ranking value based on the original point value. This produces a density map that shows different shades of color depending on the value in the cell. So areas of high solar density are in dark red, whereas areas of low density are in light blue, for example.

The next step is to assess the proximity to an electric line. The criterion is that the closer the proposed location of the solar system is to the line, the better. The base data for this analysis is a set of point measurements, with each point representing the distance to the closest electric line. The density analysis tool converts this data into a raster data set, in which each cell or pixel has a uniform value, based on the distance to the nearest electric line. In this case, where a low distance is desired, the raster data set would be reclassified to the same scale as the solar density map, with short distances having a high value, whereas long distances are assigned a low value. This produces a proximity area map that shows certain cells that are close to electric lines as darker and other cells that are farther away from electric lines as lighter.

The final analysis is to create a third map that combines the pixel values of the solar density map with the pixel values from the proximity map. The GIS tool that performs this is the overlay solver or tool. This tool will be used extensively throughout this book. In a weighted analysis, the person performing the analysis might decide that solar density is more important than distance to an electric line. In this case, the weighted analysis gives greater weight to the more important factor before combining the values of each cell. The resultant data set consists of a series of cells, with each cell having the combined weighted value. The resultant GIS map shows areas that have the highest solar density that are the closest to an electric line in dark colors compared to other areas in lighter colors. The resultant map is often called a mashup. So this mashup fine-tunes the analysis and, in effect, narrows down the search for the best place to put a solar system.

Say, for example, the solar developers also had a map of certain parcels of land for sale or lease at different prices, the developer could create a raster map where each cell represents the lease value of the parcel. They would give pixels in parcels not for sale a 0 value. This map could be reclassified on a scale where the lowest lease cost for a parcel has pixel values that are the highest value, say, normalized on a scale from 0 to 100. In the same way, the three data sets (the solar density, the distance to electric lines, and the lease cost) can be combined to give yet another map where the dark areas are those with high solar densities, nearest an electric line, have parcels for sale, and at the lowest price. Again, the analyst can decide to weigh lease costs lower than solar density, since it might be more important that the land be available than the price. Figure 2.10 outlines the steps in the analysis.

The most important aspect of site selection is narrowing down all the factors that go into obtaining the right site. The more factors that are involved in the analysis, the better the selection. In a solar application, slope, soil stability, history of cloud cover, rainfall, and dust density are all factors that should be included into the analysis. Note that the site selection spatial analysis involves the following:

- Converting all factors into simple raster images based on known or measured data;
- Converting each pixel value of the raster data set into a common scale so that the values in each cell can be added together;
- Weighing each data set if necessary;

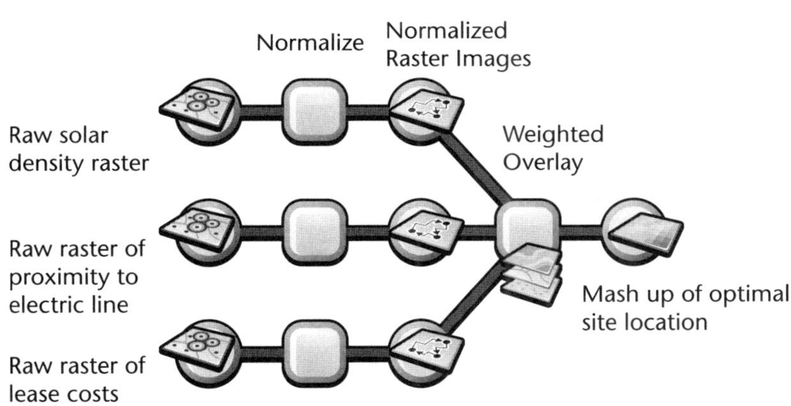

Figure 2.10 Spatial analysis for solar system site selection.

- Adding the pixels that occupy the same exact pixel location to form the output map.

The end result is a heat map, where the hot spot or darkest area represents the most optimal result.

During the exploration phase, developers of power sources scour the landscape to determine the best place to develop a plant. That process involves land availability, suitability, and economic evaluation. These factors are related to location and can be expressed separately as a GIS dataset. The analysis then involves performing a spatial analysis to uncover the best location that balances all the factors involved. The result is not a single answer, but a series of areas that represent the various grading of all the factors represented as a simple-to-understand map: the dark red areas, for example, represent the best locations; the light blue areas represent the worse locations, with other areas in between colors and densities

The factors will be quite different depending on the type of plant being considered. For example, a coal plant is highly dependent on a transportation system or needs to be near a coal mine of a certain type. A nuclear plant has to have the ability to store spent fuel, so proximity to sensitive habitat may be an issue. A natural gas–fired plant needs to be near a high-pressure gas pipeline. A wind farm needs to be in an area of high wind density. All plants need to be near existing electric infrastructure as much as possible. However, while the factors for each plant are quite different, the process of site selection using GIS is the same regardless of the type of plant.

Development

The next phase in the development of any plant involves the production of the plant itself. Once the general location of the plant is determined, the next steps involve the following:

- *Planning*—here GIS is used for site development, plant logistics, drainage determination, and permitting.
- *Surveying*—GIS is used to craft the legal rights and the exact determination of the land for the plant and surrounding support properties.
- *Environmental assessment and mitigation*—GIS is the tool of choice to evaluate the impacts of the plant on the environment, including noise, pollution, possible contamination, and control.
- *Permitting*—GIS is used to submit the documentation of the prior steps and as a visualization tool to communicate to the permitting authorities exactly the scope and impact of the project.
- *Engineering and design*—CAD is typically used during this phase for the development of detailed construction plans, but GIS can also be used as a master management system to keep track of all issues and factors that arise during the design process, including back references to prior stages.
- *Construction*—GIS is often used for major construction projects in conjunction with construction scheduling tools for logistics and organization of documentation.

- *Commissioning*—GIS can use be to organize the data in preparation for turning the plant over to the people who will operate the plant. This will be the phase that organizes the transition from a project to an operational facility. The location of all critical features of the plant will be captured in the GIS.
- *As-built documentation*—this is the process where the final documentation is completed for the plant. Changes due to field conditions and scope changes are incorporated into the final set of construction records. It is the lack of follow through on this phase where information can get lost and create costly problems later during operations. The GIS can provide the bridge from the state of what was originally engineered to what was actually built.

Figure 2.11 shows the typical lifecycle of plant development.

Managing Support Services

There are common process that must be considered when both running a plant and building one. These services involve supply chain, readiness for weather events, security, facilities management, ongoing environmental compliance monitoring, public relations, and safety. Most of these processes are dependent on location.

The movement of construction material, fuel, people, or vehicles represents a significant cost to the developers as well as operators of a power plant. For example, the routing of inspection personnel to hundreds of scattered wind turbines can be a significant logistics problem. GIS is widely used by delivery companies, warehouse managers, and fleet managers to optimize routes and select the best location of warehouses or spare parts depots. Chapter 8 contains more information on routing and supply chain.

All plants, regardless of whether they are in operation or under construction, must be ready for adverse weather events. GIS coupled with weather services can accurately predict where weather events will occur. Plant operators can perform risk assessments of their facilities to see just how vulnerable parts of the plant are to certain events, such as tornadoes, earthquakes, or landslides. With the plant modeled in GIS along with risk assessments, plant operators and developers can create risk scenarios and then create mitigation plans.

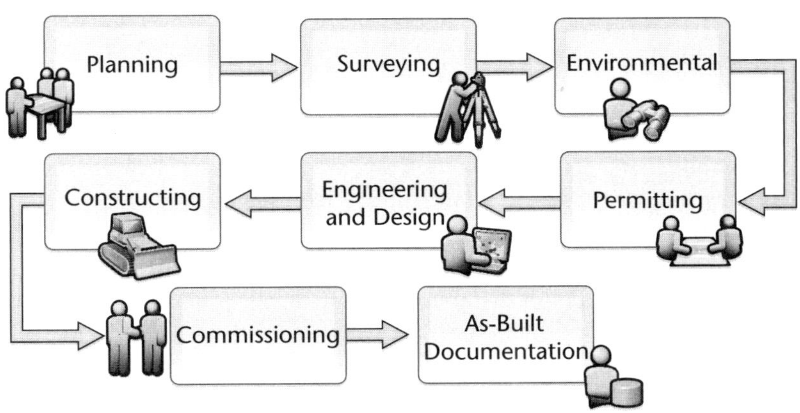

Figure 2.11 Typical plant lifecycle development.

A GIS spatial risk model is quite similar to the siting model. Data for each risk factor, such as the areas of highest incidents of wild fires, are converted into a density map and then normalized over a consistent scale. Each risk factor represents a raster layer. For example, a wildfire prediction density map is one layer, a tornado incidence map is another layer, and flood zone is yet another layer. Then by performing a weighted overlay process, a composite map that combines all the risk factors into a single density map is created. A heat map then shows the areas with the highest risk. Again the process is similar to the siting example, but the results serve quite a different purpose.

Plant physical security is critical. GIS can help with placement of cameras and sensors for optimal coverage. It's not uncommon for operators of large plants that span many acres to create a GIS dashboard that illustrates the boundaries of the plant, location of security camera, and access areas.

In addition to the actual plant equipment, power plants have a number of support buildings, warehouses, garages, sheds, storage fields, miles of roads, power and lighting systems, water distribution, waste and storm water systems, fire suppression systems, and even housing. All of these facilities need tending to. GIS can provide a way to organize the activities to support these facilities as part of an overall facilities management function.

Environmental compliance is key to the successful and continuous operation of a power plant as well as the successful completion of a plant under construction. All aspects of the environment need to be modeled, including wetlands, habitat, burial grounds, sacred lands, bird migration patterns, aquifers, and the air quality. Given the enormous impact a power plant can have on its surroundings, it's critical that operators and developers have a solid handle on the surroundings and potential hazards. Many electric devices are oil filled. A simple rupture of a power transformer could create a major environmental event.

A power plant has a significant impact on the community. As such, the operators must assure the public that they are doing everything to protect the community and all the people who work at the plant. GIS can model areas of personnel risk and monitor if people wander into areas of high risk. Operators need to convey their risk mitigation plans to the community in a way that makes sense to them. A GIS map is one of the best ways to do this.

Managing Information

Utilities routinely stored their old design drawings in a secured vault. While the old drawing vault was inconvenient, at least it was the single place that people could go to find the history of why something was done this way or where a specific permit was granted or what the results of an deep sea survey were.

Today, that information can get fragmented, often appearing in different forms, like spreadsheets, word processing documents, and private databases. One way to keep a better handle on the history of the development of the plant and to keep track of the current facilities (including any new additions to the plant) is to build an energy supply information model. The GIS is the framework, since location is so important to the information in the plant. The GIS can provide an easy guide for other related digital data and information systems, like CAD drawings, document management files, and work management systems. An information model consists of (from Appendix A):

1. *The data model*—exactly what data the GIS directly manages;
2. *Information products*—what the users will see;
3. *Workflows*—exactly what the GIS does to the data to produce what the user will see.

Figure 2.12 details the elements of the information model.

The Energy Supply Data Model

Each type of plant has its own data model. However, in each case, the model would include the following:

- *Asset data*—this would include a database schema of all the facilities within the power supply system;
- *Relationships*—the model would document the relationship of asset features to other asset features, such as control cables to cable trays;
- *Base maps*—the model would reference the core geography and terrain of the plant;
- *Raster data sets*—these include data structures about weather, natural disasters, and other nonasset-specific data.

The data model includes authoritative data; predictive information; measured data; and perhaps even social media information about how a community feels about a power plant.

Authoritative data about a wind power plant, for example, would include turbines, gathering cables, substations, control cables, conduit and auxiliary structures, and land features. Specific data would include physical and electric data about the turbines, such as height, weight, blade dimensions, and linkages to detailed construction information, electrical ratings, and maintenance history and inspection results. Authoritative data about other kinds of plants would include all major plant components and their ratings and history.

Predictive information for all kinds of plants would include data about past tsunami damage, flooding predictions, lightning, earthquake, community opposition,

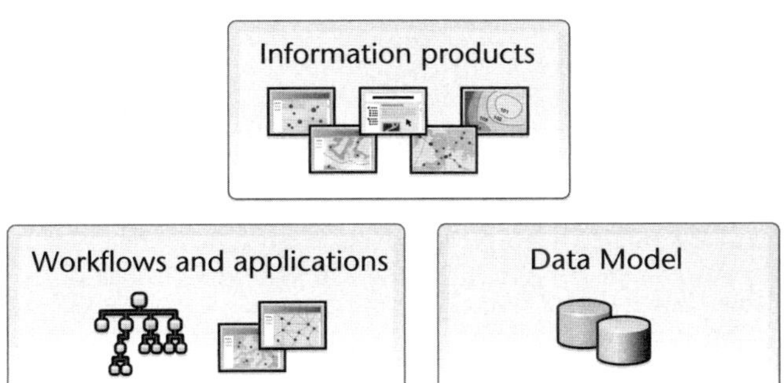

Figure 2.12 The GIS information model.

crime, and any number of data sources that relate to things that might happen and where they might occur. GIS models this data. Combining authoritative data with predictive data can gain additional insight to operators and developers.

Measured data would include real-time information from a variety of sensors. These include standard electrical sensors, like power flow, current, and voltage, to fault detection, to land movement, to ocean bottom shifts, to all kinds of real-time weather and wind direction and speed data. The vast majority of sensor data has location as a key attribute. Adding measured data to authoritative and predictive information just enriches the information product.

Finally, operators can mine social media data to provide all kinds of insight into what the community at large is thinking. They can analyze the data and display it geographically. For example, it is possible to analyze millions of tweets to determine the overall mood of the tweets. GIS can plot these moods on a map. Operators and developers can gain a better insight into their operations and future expansion plans. One of the most costly and troublesome aspects of building any substantial project is the permitting process. Using GIS authoritative, predictive, measured, and social media data gives developers a more focused insight about what to expect during all phases of plant operation and development.

Information Products

The end result of a work product, like a design, a survey, and a risk assessment, is an information product. It is the information that appears as an output of a workflow. This is often displayed in the form of a map of some kind or at least referenced in some way to a map—to a location. When designing a GIS for particular process, it's important to have a solid understanding of the elements and information that needs to be displayed. In other words, what information does the consumer of the process or workflow need to see to be able to proceed to the next process or workflow. Here are some examples within the energy supply business of what users need to see:

- Hydro plant production planning map;
- River topology showing potential flooding areas;
- Flood plain map;
- Upstream and downstream flow maps;
- Vegetated wetland maps in proximity to flooded potential;
- Evacuation maps and optimal routes;
- Facility maps for power plant;
- Security risk assessment;
- Access maps;
- Nuclear plant evacuation plans;
- Land ownership maps for lease management;
- Solar density maps;
- Wind density maps;
- Ocean floor topography;

- Shipping lanes near of shore wind plants;
- Shipwreck location plans;
- Optimal areas for wind, solar, and geothermal plants;
- Coal pile seepage plans;
- Oil spill hazardous areas;
- Geothermal resources map;
- Maintenance and inspection routing plans;
- Underground electric, water, gas, wastewater, telecommunications locations inside the plant;
- Grounding plans;
- Lightning protection plans;
- Flood management plans;
- Emergency management plans;
- Emission models and prediction plans;
- Tidal profiles;
- Gas supply pipeline risk models;
- Locations of distributed generation.

These represent just a sample of the types of information products that could be useful in the operations and development of a power plant. The information products contain data. This data is either stored directly within the GIS or is accessed by the GIS in order to produce the information product. The trend in computing today is to keep the information products relatively simple and focused on conveying only a limited set of information.

Workflows

Workflows are the part of the information model the details what the GIS does to the data, often along with input from the user to produce what the user sees—the information product. Once the information products are determined—in other words, what information does the operator or developer need to get in order to more effectively run or build a plant—then they can develop the steps needed to bring that information together in a workflow. Once the workflow and information products have been determined, then the data needed to drive the application can be gathered and modeled.

For example, the information product needed to determine the production of a hydroelectric plant is the hydroelectric plant production-planning map. This information product is needed to determine exactly what the flows will be through the plant for the next production cycle. The workflow consists of gathering land feature data, projected and actual rainfall, and measured water levels; performing a spatial analysis; converting the spatial analysis results into actual projected production values; and then producing the output.

Another example of a workflow is the determination of where to place security devices at power plant installation. The workflow is to gather data about all possible paths leading into the plant: data about the access roads, land features, hiding places, areas of high risk, and places to install cameras. Then the next step is to perform a spatial analysis of the data to determine the areas of highest risk of intrusion based on all the factors. The analysis could include an optimization of costs compared to the risks. The information product is a map showing areas of highest risk and the proposed locations of security cameras, motion detectors, and any other detection devices.

A workflow to determine the risk of damage of a wind power plant due to earthquake can be performed by gathering data about earthquake fault lines, data layers of predictions of various earthquake intensities, slope data, and structural data of the turbines. The workflow involves modeling various earthquake scenarios, combining the scenarios into various GIS layers, and performing a weighed spatial overlay to determine which turbines are most at risk during various earthquake intensities and locations.

A popular workflow is the one that shows the solar electric potential of an urban area. The workflow involves converting raster images of rooftops into numerical areas, gathering solar density data, calculating shadows and sun directions, and determining the degree of solar production from each rooftop. The information product is a GIS map that shows each rooftop in a different color that illustrates the relative merits of installing a solar system on the roof.

Other workflows can be developed that relate to important community issues, such as showing a map of where people are most sensitive to the expansion of the plant. This workflow could include the results of a mailing campaign, an online survey, and data from news services, as well as real-time social networking feeds. The information product could show a map of where community activism is highest. This may result in a different long-term expansion plan.

So each information product will have its own workflow. As noted, once the information product and workflows are determined, then the data to support the workflow can be determined.

Location Matters in Energy Supply

The days of highly regulated prices for the generation of electricity are coming to an end. Power generation is a competitive business. As such, the developers and operators of power plants need all the tools they can get to gain a competitive advantage. Since nearly every aspect of power generation has something to do with where things are, GIS is a vital tool to help them keep the plants running well, keep employees safe and productive, make money, and stay out of trouble. GIS helps power generation operators and developers manage the disparate data sources in a more effective way by organizing the data by location, helping them see patterns and connections that they could not see, and distributing and disseminating information in a simple-to-understand way, the way in fact that people have been doing for centuries: showing the data on a map.

CHAPTER 3

Electric Transmission and GIS

Thomas Edison created the first electric power system when his Pearl Street Station in New York City went live in 1882. The system included a direct current (dc) generator and underground cables connecting less than 100 customers. Edison chose the voltage of the system to be around 100 volts as a way to balance the effectiveness of the newly perfected light bulb and having a voltage not too high to be too dangerous. In this original system there was no notion of transmission or distribution. As we now know, the length of a wire from the generator to the customer was limited to about a half a mile. The only way for Edison to grow the system would be to add more and more generators separated by about a mile apart. In effect, Edison would have to create a network of distributed generation to be able to expand the system.

At about the same time George Westinghouse and his associate Nicola Tesla invented alternating current, which permitted the electric system to scale. The key of course, was the transformer, which allowed the fledgling utilities to raise voltages for transmission of power over long distances. Transformers then could step down the voltage for safe delivery into homes and businesses.

Location Is Essential to Transmission

Even in these early days of the development of the power system, location was important, and most early utility operators mapped their systems. Figure 3.1 shows an early map of Edison's original service territory. Note that electric utilities have used maps from the very beginning of the industry.

Along with the invention of alternating current and the transformer, Westinghouse and Tesla invented and created the electric transmission system. The notion of the transmission system emerged over the years to mean high-voltage current carrying equipment, supporting structures, and auxiliary equipment that delivered bulk electricity. The terms the grid, the high-voltage grid, and the bulk power supply system all are ways of describing the electric transmission system.

Today, in the U.S., the official definition of the bulk supply system is any electrcial network with a voltage level of 100,000V or higher. Most other countries follow this general rule. Subtransmission lines that operate below 100,000V, such as at 69kV, are really just older transmission lines.

This chapter deals with the electric transmission and how the operations and system development of the transmission system depends heavily on location and of course GIS.

Figure 3.1 Edison's service territory around Pearl Street Station. (Source: Smithsonian.)

Transmission Lines

There are three types of transmission line construction: overhead, underground with solid dielectric cables, and underground pipe type cable. Underground transmission lines consist of high-voltage cables that utilities bury directly in the ground. The last type, pipe type cable construction, consists of insulated high-voltage cables contained within metal pipes. Equipment pressurizes the insulating oil and pumps it through the pipes to increase the cooling of the cables. The system cools the oil at each end of the cable by forcing the oil through heat exchangers. The power delivery capability of this type of underground construction is higher than direct buried solid dielectric for the same conductor sizes due to the additional cooling provided. Here the location of the pipes are of course critical, but also the proximity to sensitive environmental areas is important, should a pipe leak or rupture sending oil into these sensitive areas. So not only is location of the pipe and its assets important, but proximity is just as important.

Transmission Substations

The transmission from the power plant to the transmission or bulk power system occurs at the generator substation. Today's modern generators utilize voltages in the medium voltage range. Typical values range from 12kV to 24kV. Most modern power plants link each generating unit with its own step up transformer that converts the generator voltage from the medium voltage range to the transmission voltage range. The number of generating unit transformers at the plant depends on how many generating units a power plant has, typically one for each generator. Older power plants may synchronize several smaller generators onto a single generator bus, but a one-for-one generator-to-unit transformer is more common.

The generator substation then consists of the high-voltage feeds from each of the unit transformers often into a sophisticated bus arrangement. Figure 3.2 illustrates a typical generator substation schematic. Exiting the substation are one or more transmission lines, each one protected by one or more high-voltage circuit breakers. Thus, the generator substation marks one of the transition points between generation and transmission. Equipment measures the exact values of power, voltage, and current at the terminals of the unit transformers. These measurements provide insight into both the business and technical aspects of the transmission system. These values help enhance the visualization of the system by the GIS.

The transition from bulk power delivery to local power delivery occurs at the many high-voltage (HV) to medium-voltage (MV) substations. There are a number of different configurations of bus work and medium-voltage feeders of these substations. The substation taps off the transmission system. Substation transformers step down the voltage from high voltage to a medium voltage. The transformers in turn feed the medium-voltage switchgear. The medium-voltage switchgear supplies the distribution feeders. These substations define termination points of the transmission system. Figure 3.3 illustrates a simple one-line schematic diagram of an HV/MV substation. Utilities locate metering equipment here to determine how much power the distribution system consumes. These metered values provide additional business and technical insight to the behavior of the transmission system and are valuable for the transmission GIS.

Other important components of the transmission system are the high-voltage system switching substations. These are the HV/HV substations. They serve as a way of interconnecting several transmission lines to facilitate switching and line protection. Some substations also provide a transition between two (or more) high-voltage transmission voltages. For example, a utility may have built and operated transmission lines at 230 kV. Later as the technology matured, they built and operated lines at 345 kV. An autotransformer connects two different high-voltage systems. The autotransformer consists of a single winding for the higher voltage and a tap off point for the lower voltage level. The switching substation provides the

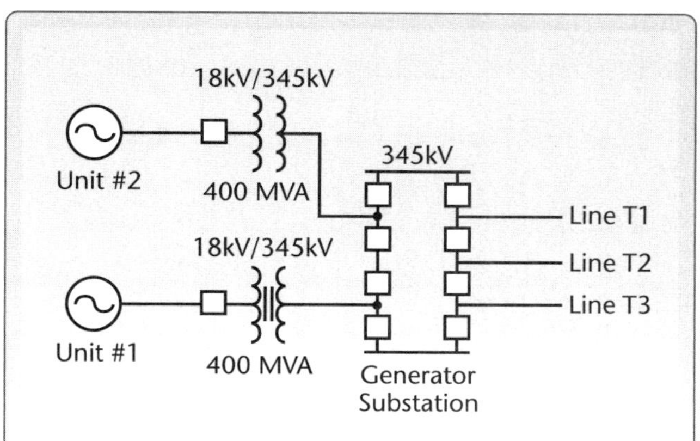

Figure 3.2 One-line schematic of a generator substation.

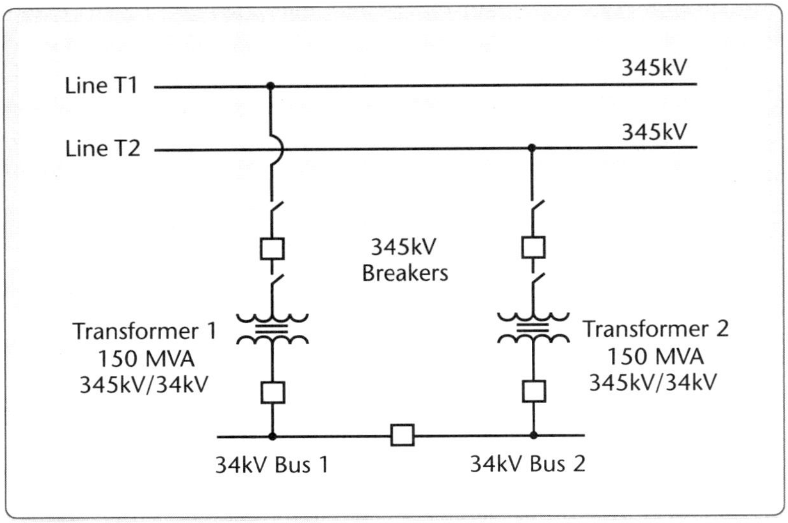

Figure 3.3 One-line schematic of a HV/MV substation.

breakers, protection, and measurement schemes to be able to manage these kinds of connections. Figure 3.4 shows a one-line schematic of an HV/HV switching substation in which the two high-voltage levels are connected with an autotransformer.

Alternating current is not without its difficulties. In an alternating current system, the waveform for the voltage and currents follow the patterns of trigonometric functions. What regularly happens is that the voltage and the current get out of phase. The more out of phase the voltage and the current, the less efficient the system is in delivering power. In fact, the cosine of this difference in phase between the voltage and the current is the power factor. A power factor of one, which represents a phase angle of zero (the cosine of zero is one) means that the voltage and current are exactly in phase. Utilities strive to keep the power factor as high as possible. In an ac system, voltage has two mathematical components, the voltage value (the voltage magnitude). A simple volt meter can measure this. The

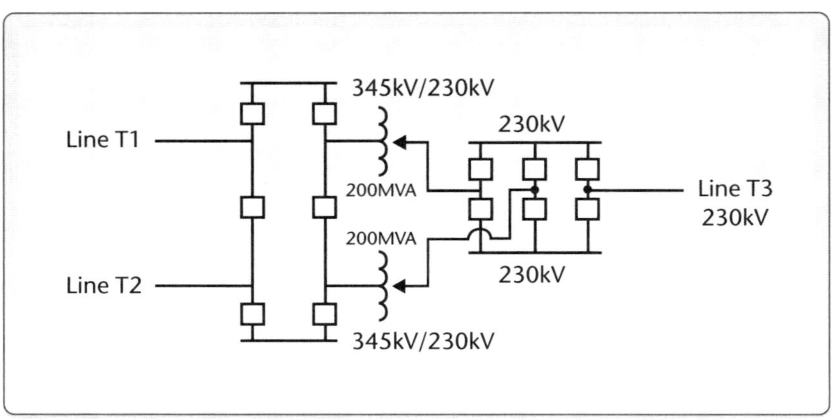

Figure 3.4 One-line schematic of a high-voltage switching substation with autotransformers.

other component is the relative phase angle from some reference angle. This is not so easy to measure. The difference in the phase angle has a strong influence on the amount of power that flows. Thus, utilities use phase-shifting transformers, which have the ability to increase or decrease the angle difference and thus control power flow. These represent an additional component of the transmission system. These phase-shifting transformers are large and most always located within the transmission switching stations. Phase shifting transformers make up an important part of the transmission GIS.

A new concept in the ongoing problem of actually measuring the phase angle is the so called phasor measurement unit or synchrophasor. Since voltages and currents vary with time according to trigonometric rules, these devices synchronize the timing with a GPS clock. The synchrophasor system measures, transmits, collects, and analyzes the data to produce the phase angle at every measured point in the system. The result is the relative voltage phase angle at each point where the measurement has taken place. Prior to these systems, engineers had to perform complex network analysis to determine phase angles. The synchophasor system is highly dependent on location and GPS, so utilities will need to include this system in a transmission GIS.

HVDC

Today, Edison would certainly be vindicated. Direct current actually is quite capable of transmitting power over long distances but at high voltages. Standard ac transformers convert the voltage to high-voltage ac, and then power rectifiers convert the high-voltage ac to high-voltage dc (HVDC). At the other end of the long transmission line, the voltage is then reconverted back to ac using power inverters.

Unlike the ac system, there are no phase angles, power factors, or time-dependent variables to worry about. HVDC lines have the advantage of isolating disturbances along the transmission system. Whenever ac transmission grids are interconnected, events such as a major failure of a line or a sudden loss of a number of generator units create a disturbance throughout the system and could cause the system to become unstable. Just prior to the 1965 Northeast blackout in the U.S., the electric system became unstable. The instability is largely a result of failure of an uncontrolled variation in the frequency of the network. With HVDC, there are no frequencies involved, so the HVDC lines prevent the instability of frequencies to propagate. In effect, an HVDC line provides a way of transporting large blocks of power from one grid to another while filtering out disturbances.

Transmission System Components and GIS Data Model

While utilities will certainly maintain the information about each of the various parts and components of the elements of the transmission system, the location, type of part, and condition will play a major role in how well the system functions.

The parts of the transmission system are as follows:

1. Generation substation, which receives power from the power plants;
2. Transmission lines, which deliver the bulk power to the HV/MV substations;

3. HV/MV substations, which deliver power from the transmission system to the distribution system (or large industrial loads);
4. Switching stations, which provide a means to switch and configure an integrated system of transmission lines, and often includes autotransformers and phase-shifting transformers;
5. HVDC stations, which include power electronics and the ac power transformers and associated protection and accessories;
6. Sychrophasors, or measurement systems that capture the relative voltage phase angles;
7. Measurement systems at the various transmission transition points and changes of ownership;
8. Structures;
9. Transmission corridors and parcels.

From strictly an inventory perspective, utilities need to know where each of these components are, their relationship to one another, and their surroundings. This is what the GIS does. In addition, the GIS enhances the understanding of the system by spatial data management (where things are) analysis (what the data means in a spatial context), data enrichment (adding more meaning to the data by combining spatial information and analysis), and visualization (providing a framework for decision making).

A System in Transition

As primitive as it was, looking back at Edison's original delivery system, its purpose was clear—deliver power from the source, the generator, to the load. For the greater part of the twentieth century, the transmission system's function was the same—to provide a means of delivering energy from generators to loads. Since late in the last century, the transmission system now is an open access vehicle for a competitive generation and wheeling marketplace. The transmission system has evolved into a vehicle for financial transactions as well as a vehicle for delivering power. In the past, the interruption or failure of a transmission system component prior to the open market could have resulted in outage. Today, a failure of the transmission system component results in both an interruption in power and a disruption in the market. Thus, the accurate understanding of the system using GIS is more critical today than ever before.

The Business of Electric Transmission

The industry is in a transition regarding electric transmission. As noted in Chapter 2, electric generation and energy supply are moving to full competition. In many countries and in a several states in the U.S., governments have required the industry to unbundle into its four component parts—generation, transmission, distribution, and retail. In some countries, the communities have created a single transmission utility. Nonetheless, the transmission business is still a regulated monopoly. The challenges and concerns from a business and operating perspective have not changed much from the vertically integrated utility days. So, whether the transmission is a separate company, a state-owned transmission agency, or still part of a vertical electric company, the work, processes, and organization of a transmission system are the same. Figure 3.5 illustrates the main processes of a transmission business.

In addition to the actual transmission operators and owners, there are a number of regulating authorities that actually coordinate and in some cases take control of the transmission system. In the U.S., for example, there are independent system operators, like the California Independent System Operator (CALISO). There are regional transmission operators (RTOs) such as the PJM Interconnections. RTOs are similar to ISOs, except the U.S. government has required that they comply with the definition of an RTO. Overseeing the reliability of the grid is the North American Electric Reliability Council (NERC). NERC has legal authority to direct transmission operations regarding reliability. In essence, the owner and operator of a transmission system has to coordinate its activities with a power pool, an ISO or an RTO, as well as NERC. By coordination, the owner must seek permission to perform certain actions, like switch a line out for maintenance.

What's critical for the smooth operation of all these entities is communication and collaboration. The GIS provides one additional way of making sure that everyone involved in the transmission system from the line worker in the field doing an inspection or maintenance activity to the reliability assessors at one of the NERC reliability centers has access to the same information. Collaboration among the various parties making up the transmission system can get complicated.

What Does GIS Have to Do with Transmission?

Transmission is strategic. In August of 2003, the United States experienced a major blackout. The trigger event was when two small trees came too close to one of several transmission lines interconnecting the Canadian grid with the U.S. grid in New York state. Under normal situations, this should not have caused a major blackout,

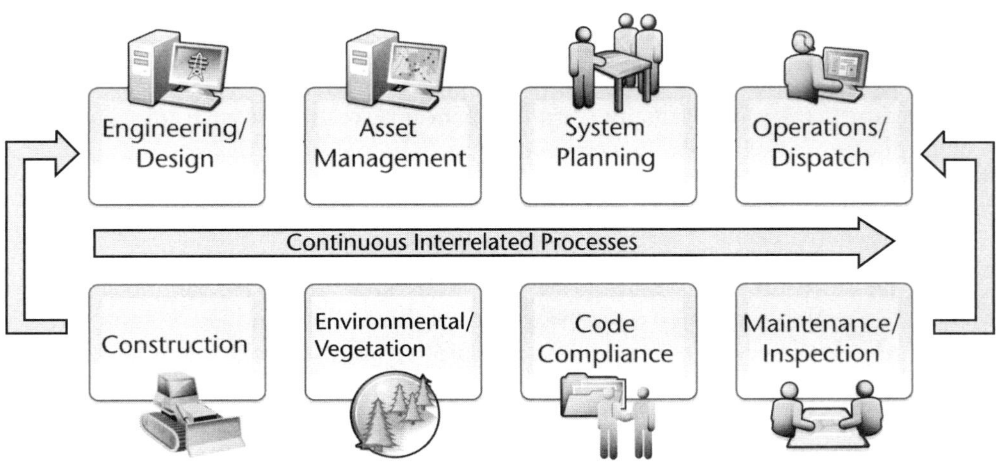

Figure 3.5 The common processes of a transmission business unit.

but some technology did not function properly and some of the operators did not take appropriate action. When referring to this particular event, people say the three "T's," namely, trees, training, and technology, caused the blackout. The lack of solid situational awareness, tools for collaboration, and up-to-date data probably contributed just as much to the blackout.

GIS is about creating a means to collect information on a common visualization platform, namely, a map, and tell a story, often a story of what's actually going on right now. During major events (like the 2003 blackout) the community of operators regularly do not have a single picture of the event, nor do they have a spatially oriented risk profile of the transmission system to see where vulnerabilities might lie (e.g., a map of the transmission segments where vegetation management had not occurred in recent years, overlaid with transmission lines that were identified as critical tie lines).

All transmission entities from operators to power pools to RTOs perform complex network analysis, like power flows, short circuit analysis, stability analysis, and state estimation. These analyses, while essential, only tell part of the story. GIS provides the spatial context to the network analysis. Today's modern GIS can easily consume data from these analysis systems. GIS can provide a single view of data from a variety of sources into a common operating picture or dashboard.

What is involved in transmission and what does GIS have to do with it? This remaining part of this chapter will examine in more detail how GIS impacts the following:

- Operations;
- System development;
- Support services.

What Transmission Operators Worry About

A transmission operator like the power plant operator has four important stakeholders (shareholder, customer, employee, and community) and essentially four missions. Paraphrasing from the preface and the former CEO of a power company, transmission operators need to make money, keep their customers happy, keep employees safe and productive, and of course stay out of trouble.

Make Money

The transmission operator's mission is to make money for the owners—to keep the power flowing. The system is very complicated to run. The owner invested in the system. The end goal is push as much power through the existing infrastructure as the operator can possibly do without damaging anything. If, for example, operators run the system conservatively, then they leave money on the table. If they run the system too hard, they risk damaging the investment and could lose the revenue from the power flow over the line. So having good information about the system is essential to running the system at its optimum flow.

In addition to keeping the revenue flowing, transmission operators also need to keep expenses down. GIS tools such as routing, vehicle location, optimizing expensive vegetation management, and analysis of when to replace sections of the system are critical for keeping expenses under control.

Transmission owners also want to be as effective as they can when building new transmission facilities. Building a new line can be a minefield: acquiring land, knowing what the best route is, permitting, dealing with abutters, and optimizing the costs to provide the largest capacity line for the lowest cost. The land, the line configuration (the number of turns, changes in elevation), and the potential hazards drive the cost. GIS plays a strong role in managing new infrastructure.

Keep Customers Happy

Direct customers of the transmission business units are the distribution utilities and large industrial customers who tap off the transmission system, the generators who pay the transmission company to use their lines to carry the power to the retailers, or the indirect customers—everyone else who might be impacted by a failure of the transmission system. The goal here is to keep the system running without failure to keep the power and the money flowing.

Keep Employees Safe and Productive

Safety and good information go together. The transmission system is complex and physically diverse. Certainly employees will know where the towers and lines are. However, it may be difficult for employees to access the transmission right of way due to flooding, hazards, fires, earthquakes, or even terrorist activities. Much of this information is geographic. For example, in the case of a wildfire, utility workers inspecting, maintaining, or troubleshooting the transmission may find a wildfire has trapped them. Having a solid geographic reference for escape routes, the fire path, and the wind patterns could save lives and prevent disasters.

Giving employees correct and timely information about their work, and routing them in a scientific and orderly way, improves the overall performance of the utility workers or contractors. Labor is one of largest costs for a utility. Field labor represents the bulk of that labor costs. Utilities can fine-tune the movement of people and material using GIS.

Mapping tends to be a language that everyone understands. It was one of the earliest communication tools of society. GIS allows groups, teams, and individuals to communicate problems, risks, status and locations for the purpose of collaboration. With GIS readily available on desktops, the web, smart phones, and tablets, utilities can collaborate using the ancient tool of mapping to better support the business. Giving employees tools that better foster communication and collaboration results in fewer missteps and an overall safer and more effective work environment.

Stay Out of Trouble

On July 31, 2012, one tenth of the world's population suffered a blackout. It was the most severe blackout in history, 130 years after Edison switched on the first electric grid. The problem was the failure and collapse of much of the bulk supply system, the transmission system. The cost of a blackout of this size to society is enormous. The cost to a utility can be catastrophic in real dollars and in diminished reputation. GIS cannot make the grid more robust or add a single kw to generating capacity or improve the voltage. What GIS can do is provide a framework for better insight into what is happening and what could happen.

Managing Transmission System Operations

The main drivers of the transmission system operator are as follows:

- Keep the power flowing at all times to the customers.
- Run the system as efficiently as possible.
- Prevent accidents.
- Comply with all regulations.

GIS in the Control Room

The people huddled in the control room operate the transmission system on a second-by-second basis. The transmission dispatchers watch the transmission system carefully, monitoring its every movement. When a sensor buried deep within a power transformer at a switching substation sends an alert that the nitrogen levels are too high, the transmission system operator or dispatcher makes a decision. That decision could involve a switching action like taking the transformer out of service and rerouting power along a different path, creating a work order for the maintenance crews, doing nothing and waiting, or a variety of other actions.

The main tool of the dispatcher is the supervisory control and data acquisition (SCADA) system. It allows dispatchers to remotely control the transmission system from within the control room or energy management center (EMC). Nearly all transmission SCADA systems consist of the control computers located at the EMC and remote terminal units (RTU) at each of the substations and other locations throughout the transmission system. The RTUs marshal all alerts and transmit those alerts to SCADA. In the example of the nitrogen gas alarm, the RTU sends the signal to SCADA indicating that a particular transformer had an alarm. Depending on how comprehensive the SCADA system is, the alarm could be that something is amiss at the substation, that something is amiss at a particular transformer, or that there is a high nitrogen alarm at transformer 1B. SCADA receives this information and displays it to the operator.

The vast majority of visualization tools in SCADA systems are schematic. They show detailed substation schematics, like those in Figure 3.4. They show the transmission lines that interconnect the substations but do not show the physical representation of the transmission lines or substations. For example, if a fault occurs on a 25-mile section of a transmission line, SCADA creates an alarm that a fault has occurred and it will show which breakers tripped. SCADA has no mechanism to show the operators where the fault is or what conditions might be at that location.

GIS in the Field

The next step for the dispatcher is to send a troubleshooter to the area of the transmission system to locate the fault. The dispatcher may get other clues as to where the problem might be, such as a police report of a fire near the line or a report of a loud noise or flash or plane crash or some other event that may (or may not) help the troubleshooter narrow the search for the problem. Of course, if the fault happens to be in an area that is inaccessible by a truck, the troubleshooter would then have to perform a foot patrol or obtain a snowmobile or an offroad vehicle. The point is that given that the line is 25 miles long, the process to actually locate the fault could take several hours.

A loss of an element of the transmission system results in a loss of capacity, which could result in a loss of revenue if the line results in outage. It is thus critical for the operators to assess the damage, put processes in place to fix the problem, and get the line back in service as quickly as possible. One line out of service increases the risk of the failure of the entire transmission system. If, for example, another line has a fault, this again increases the risk of catastrophic failure of the system. At some point, utilities know that there is a tipping point in which the loss of one more component could cause the system to go unstable. Utilities know that they must never get to that tipping point. What they need to do before they ever get to that point is to shed (drop) load. Load shedding does what every transmission operator and owner does not want to do, shut the power off for some of the customers. Sometimes the term for shedding customers' loads is *rolling blackouts*. Whatever the term, it is a bad thing, but necessary to preserve grid operations. Utilities perform contingency analyses and study the impact of losing one or more components to see exactly what conditions would need to exist to get to that tipping point.

While SCADA doesn't know exactly where the fault is, parts of the utilities' automation systems and other available information provide clues. The GIS can pull together these clues to help pinpoint the location of the fault and help assess what the conditions are near the fault so that repair crews don't have to guess what to bring to the fault site. For example, when a fault occurs on a transmission line, it generates a large volume of current. The control system reads the fault current value and takes action. A breaker opens, clearing the fault. After the breaker opens, no more current flows. Impedance relays accurately measure the volume of fault current and can calculate distance from the breaker to the fault. Since SCADA doesn't capture the location of the lines, it can't provide much help as to exactly where the fault is. However, the transmission system GIS can accurately locate the position of the fault based on information from the impedance relays. This gives troubleshooters and repair crews insight well before they have to go to the field to investigate. This information helps crews find the shortest path to the problems and gives them the nearest access roads.

The GIS helps to add additional information to the restoration and fault location effort. If there were other clues, like someone saw a fire and called the police, the GIS could display their location. The GIS then would provide a means to take authoritative automation data, as well as other information, such as fire location, to pinpoint the location of the fault. The GIS can also help determine the cause of the fault, by bringing in data from additional outside sources and systems.

A major cause of a transmission line fault is lightning. Most transmission operators have a lightning-monitoring system or service. Merging data from the lightning system into the GIS could tell if strong lightning was in the area. If the utility had a feed from the forest fire agency and discovered there was a wildfire in the region, the utility would know that the line may be inaccessible and dangerous to try to repair. Figure 3.6 shows a GIS map with real-time lightning strike data.

Understanding the Variables

The GIS does not replace the SCADA system in any way. It provides the missing spatial context and facilitates the workflow for transmission line repair.

Transmission operators regularly perform offline network analysis to help them understand the nature of the system. This network analysis simulates what might happen during a contingency or during high load levels. The result of the analysis is a listing of voltage (magnitude and phase angle) at each junction in the network. The GIS can consume this data to give operators a sense of where issues might occur, not from a schematic perspective but from a spatial one.

Compliance

Another significant problem for transmission operators is to make sure that the system is compliant with regulations. Most regulators have the same goals as the transmission operators themselves: keep the power flowing. So they reward utilities that do that and punish utilities that do not. A key element of compliance is the strict maintenance of good information. In the United States, NERC issued an alert that indicated that they believed that some transmission operators did not have accurate information about the location of their towers. They ordered the transmission operators to do whatever was necessary to assure accurate location of towers. They suggested that operators use light detection and ranging (LiDAR) to do this.

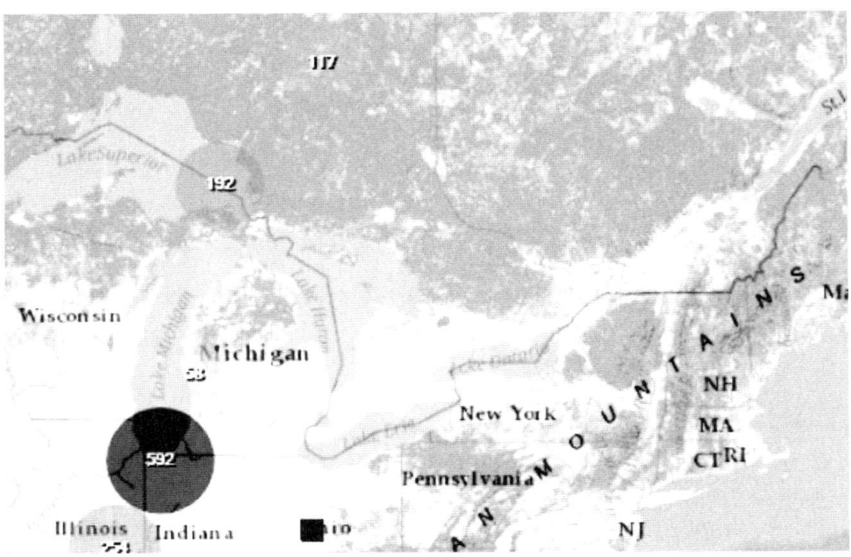

Figure 3.6 GIS map showing lightning strike data. (Source: Esri et al.)

LiDAR uses laser technology to send signals from an aircraft to the ground. Each returned signal represents a data point. The round-trip time of the signal determines the distance from the source. So LiDAR is a very accurate way of measuring the location of devices, land elevation, tower location, and tree canopy. The output of LiDAR is millions of points. LiDAR providers can characterize sets of points, called point clouds, into specific categories, like ground elevations, trees, and transmission towers and lines. GIS can display the characterized point clouds to determine additional information like any encroachment or clearance problem for the lines. LiDAR data can enhance GIS data by feature data recognition and feature data extraction. Most LiDAR data sets also come with detailed imagery of the areas as well, which can be incorporated into the GIS data.

For example, if operators do not have an exact representation of the ground elevation under a segment of line and do not know exactly where the towers are, or what the weather is, or wind speed, or age of the conductor, or the last time someone trimmed the trees, operators will of course make assumptions—conservative ones or risky ones. With a solid GIS model of the transmission system, including the ground clearances and good facilities data, the operators can make the right decision about how hard to push the line. Figure 3.7 shows a simplified profile of a transmission line with the data as the operator believes is correct. Figure 3.8 shows the line with the exact locations, but the operator is not aware of this real situation. When a line is heavily loaded, the conductors sag. Add wind, and the conductors sway. The degree of sag depends on a number of factors, but loading is certainly the most critical and controllable factor. Note the location of the knoll under the line. Since the operator does not know the exact location of the knoll in relationship to the towers and the line conductors, the operator does not know the maximum sag that would be allowable in this line section. In the first case, the operator will be too optimistic in rating the line based on a misunderstanding of the clearance of the sagged conductor to the ground.

While SCADA and substation automation systems provide insight into what is actually going on the transmission system, GIS with additional data sets from LiDAR gives the operators the exact location of their facilities. Of course, this will assure the regulators that the operators will make the right decisions about how

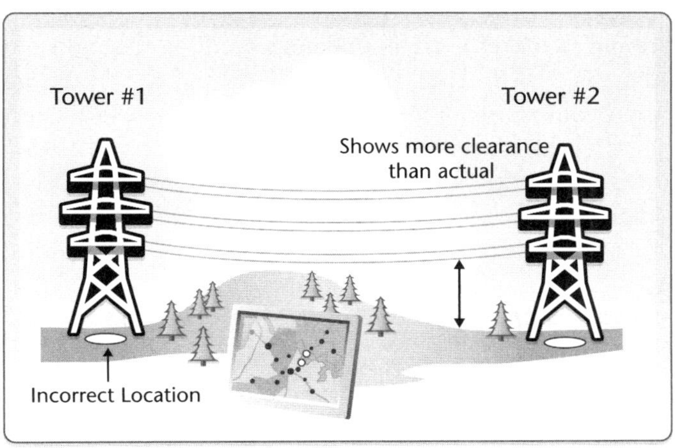

Figure 3.7 Transmission line profile showing information that the operators have.

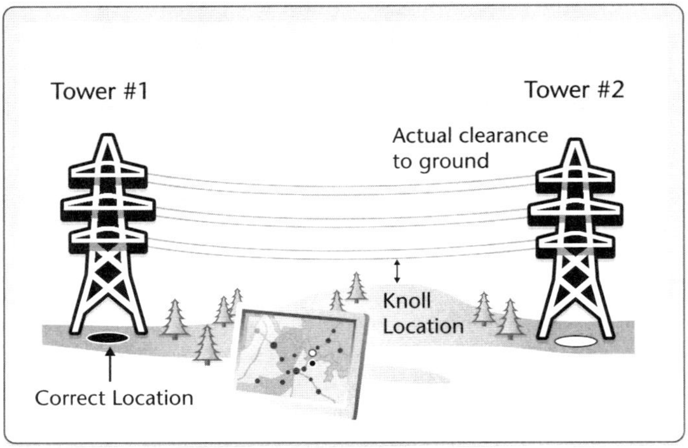

Figure 3.8 The same section of line from Figure 3.7, showing exact locations of the tower and knoll.

hard to push the transmission lines. Further, it gives the dispatchers the tools to minimize problems along the transmission lines.

Vegetation Is the Transmission Line's Enemy

Failure to manage the vegetation of an overhead transmission line invariability leads to failure or flashover of a transmission line. The traditional way to manage vegetation along a transmission right of way is to schedule the clearing or trimming of the right of way on a multiyear basis. So vegetation management planners would assign sections of the right of way to be trimmed or cleared based strictly on a frequency basis. For example, section 5 of right of way 7 would be trimmed or cleared every five years. Every section of every right of way, then would trimmed every five years on a rotating basis. The problem with this approach is that it does not take into account the natural variations in the vegetation along the sections of the right of way. It may be better to trim or clear some sections (or parts of sections) more aggressively and others less aggressively.

When expenses are tight, utilities can be tempted to extend the trimming cycle an additional year. Yet there may be places along the right of way—even short sections—that have fast growing trees. Extending the trimming or clearing just one more year could result in a clearance violation and a flashover of a transmission line.

Planners can take into account the variations of species and the variation in factors that contribute to tree growth by using GIS. This gives utilities the ability to optimize their spending on vegetation management, while focusing their efforts on those areas that are most problematic. Utilities can also use GIS to manage the various work packages that utilities create for trimming and clearing contractors.

A simple example of a GIS analysis is to discover which areas of the transmission corridor contain fast-growing trees. Of course this is a simplistic notion of vegetation management, but it is another example of the GIS overlay analysis. Here are the steps:

- Obtain a GIS-based fast-growing tree polygon raster data set along the transmission line or right of way corridor.
- Obtain a GIS-based raster data set showing average rainfall in areas along the right of way for a period of several years.
- Create a simple raster data set showing the transmission rights of way for a particular study area. The raster data set can be easily created from the vector data set of easement parcels.
- Obtain a GIS-based raster soil permeability data set.
- Obtain other data sets that may influence tree growth (e.g., sunshine, cloud cover, insect contamination, wetland area, slopes, areas of low or high drainage, etc.)
- Normalize each data set based on a common scale (e.g., on a scale of 1 to 10).
- Establish a weighting factor for each factor.
- Perform a GIS weighted overlay analysis—this creates a new map that illustrates the most likely location where trees will grow into the transmission lines.

Often the results will be displayed along with detailed imagery of the area. Utilities can also perform "tree growth analysis" based on the factors in each of the polygons.

The resultant fast-growing tree raster data set consists of regions or polygons with each pixel having a value that represents a ranking of how fast or slow the trees in the areas grow. When displayed in the GIS, fast-growing areas are shown in dark colors while slow-growing areas are shown in light colors.

Many of the data sets would come from the local conservation commissions, the government, or an analysis of detailed imagery. If a government commission provides the data as a GIS layer, the utility could request the data set be provided electronically through a web service. This simplifies the data-gathering process.

The output product of the analysis is a series of line segments or, in the GIS terminology, a new feature showing only those rights of way that cross a fast-growing tree zone or a new feature showing each right of way segment in a different color or a different shade of the same color, with each color representing the degree of tree growth (the darker the color, the faster the tree growth in that area). The analysis creates a new line feature having the same location as the original right of way.

Figure 3.9 illustrates the simple workflow of how to overlay polygon features to simple line features.

Asset Management

Transmission operators must have a solid understanding of their assets to be able to understand whether to continue to maintain the assets as they have been doing or to consider replacements. This is the essential nature of asset management. There are a variety of tools at the transmission operators' disposal for doing sophisticated analysis to getting to this answer. They can use historical data, equipment failure rates, spending information, and algorithms that can predict when a piece of equipment

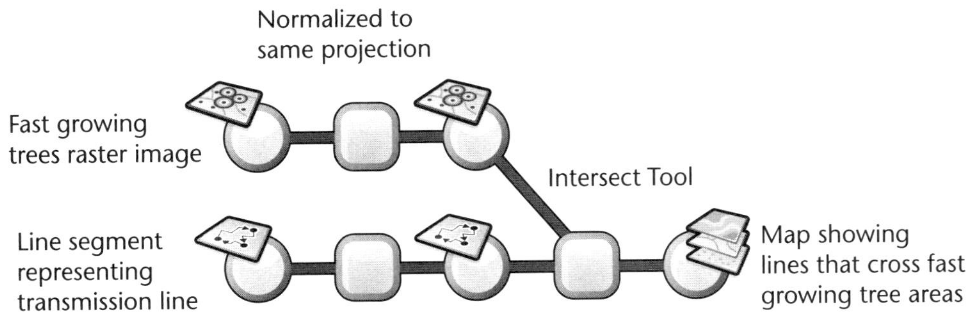

Figure 3.9 How to use overlay analysis to discover the best place to trim or treat vegetation in a transmission right of way.

is likely to fail. The one piece of data that often is the most difficult to obtain is the condition information, since that will often involve a visit to the field.

Foot and helicopter patrols are common workflows. The most common way to collect this condition information is with the use of mobile device with GIS and GPS functionality. Once the field workers capture the condition data as an attribute of the asset, then the utility can perform effective asset management.

Once the analysis is performed using asset management tools, such as an enterprise asset management (EAM) system, utilities will often want to express the results in the form of a map. However, GIS can actually add more insight into the results by including a spatial context. For example, if a section of a transmission line is deemed to require replacement, the GIS may show obstacles or difficulty factors, such as the proximity to wetlands, slopes, or landslide areas that may trigger additional study. GIS provides the inventory of the equipment, the location, and the condition and can provide additional insight near the lines that may impact the decision to repair or replace.

Risk Profiling

A question that every transmission operator would like to know is what parts of the system are more likely to fail than others. An EAM can provide some of that information, but not all. A line that is of a certain age or is subject to continuous heavy loading are clearly at higher risk, but there are other factors that are purely spatial.

Utilities use GIS to refine risk. Transmission system risk has two components: how likely the equipment is to fail and how hard it will be to repair. If a transmission line is crossing a mountain pass, regardless of its age and service duty, it will be more difficult to repair than a line running along a freeway in the middle of a city.

The concept behind risk assessment is to combine all the known data about risk into a weighted overlay analysis using the same tools that are used for vegetation management. The difference is that the data sources are different. In this case, the analysis uses the attributes of the transmission line itself from the facilities model of the GIS, like age, failure rate, and any other factor. For a simplified risk assessment for a transmission line, the process would include the following:

- Identify every transmission line segment in the study area.
- Provide inputs from EAM systems and GIS attributes for each line segment including the age, failure history, cost of maintenance, and any additional risk analysis results.
- Normalize the attribute values so they can be added numerically. For example, age can be expressed on scale of 1 to 10. An age greater than 50 years gets a 10, new gets a 1. Weightings can be used at this step if needed.
- Create a raster data set from the vector representation of the transmission line.
- Assign attribute values to the pixels along the transmission line location based on the combined transmission line attributes.
- The resultant map will then show transmission lines with various shades, based on the combined attributes. A section of line that is old and has been repaired frequently will be shown in dark red, for example.
- Obtain or create a GIS raster data set of lightning strike historic data preprocessed into lightning activity zones.
- Obtain or create a GIS raster data set of earthquake fault lines.
- Obtain or create a GIS raster data sets with landslide zones added.
- Use the vegetation management age zones—the time from when the vegetation was last attended to.
- Obtain or create a GIS raster data sets that show proximity to hazardous situations (like near municipal airports) or known high crime areas or known hunting areas.

This is just a short list. Planners can add more hazards. For example does the line segment cross a river? If so that represents an additional risk value.

The analysis involves taking each of these GIS data sources and normalizing them into a similar format. GIS tools perform this. So for example, if a lightning map is a raster map where each pixel shows a value of basic insulation level (BIL) for each location, the spatial analysis model would be convert or normalize the data to a scale of between 1 to 10, 10 being high. Likewise each factor would be converted to a similar scale, where the highest risk represents the higher value, again with 10 being the highest risk.

Once all the nonspatial and spatial data are normalized into a consistent set of spatial information all based on the same risk scale, the spatial analysis model performs a weighted overlay process plus a proximity analysis for those hazards near the line, creating a new line feature that inherits all the attributes of the other layers. Then each line segment can be visualized on a map. The GIS displays the segment with the highest risk score in the darkest color.

Other Uses of GIS for Operations

The following is a list of other transmission line workflows that use GIS:

- *Call before you dig*—digging into a high-voltage underground transmission line is deadly. The energy unleashed during a transmission short circuit is enormous. So it is critical for each operator to have precise information about where an underground cable is located. The GIS is convenient to facilitate the process of taking call before you dig requests and locating underground cables.

- *Insulation coordination*—transmission operators must make sure that the insulation ratings for the system are correct and represent the best design for possibly changing situations. Most insulation coordination software is available only as a standalone application. It is not often sold as part of an enterprise solution. So modeling the insulation systems in GIS helps to coordinate the data. Reporting the results of insulation coordination studies in GIS provide a means of giving greater insight into any insulation breakdown issues. As noted earlier, a common cause of transmission line outage is due to lightning. Flashovers from lightning are caused by a failure of the insulation to protect the line. The GIS can provide a convenient way of coordinating lightning protection with the results of insulation coordination studies.
- *Electromagnetic fields (EMF)*—in the 1990s, utilities were involved in a number of battles over the issue of EMF. The controversy centered on the claim that EMF from transmission lines caused a health risk. Every transmission line produces EMF. The strength of the EMF diminishes with distance. Utilities can use GIS to determine exactly the extent of EMF from their lines and see what lies within what levels of EMF.
- *Grounding*—deterioration of the grounding equipment can lead to failure of the insulation to protect the transmission equipment. So, having a good inventory and condition of the transmission grounding system provides the utility with better asset management and risk mitigation. The GIS could provide an analysis that studied unaccounted for flashovers and correlate that information to field inspections of grounding systems answering the questions of where have there been unexplained flashovers and where are grounding systems are in need of repair.
- *Environmental*—power transformers, breakers, pipe type cable, and other equipment are filled with large volumes of insulating oil. Should a leak or rupture occur, the transmission operator will be responsible for the cleanup of any environmental issue caused by the oil. So GIS can manage the proximity of any kind of sensitive environmental area near oil-filled equipment. The GIS can also model the various leak or rupture analysis to determine what impact that a spill might have.

Substation Management

The most common representation of the data about substations in the transmission network in GIS is the typical substation schematic or one-line diagram, like those shown in Figure 3.2, Figure 3.3, and Figure 3.4. However, in the GIS, these diagrams should be part of the connected network of the transmission system and stored at the correct location/position of the substation. In addition, each element of the substation, such as the transformers, switches, and breakers can be key reference for other data. For example, the GIS can reference the entire catalog of shop drawings, design details, wiring diagrams, and panel schematics to the particular breaker feature in the GIS. The symbol of the substation yard or building can be the locational reference for all the site-related documents, such as general layout, building, and grounding plans. The GIS can reference other auxiliary systems such as batteries, RTUs, and annunciator boards to the substation location feature.

Utilities maintain hundreds of documents for each of their substations. Having a way to organize the documents with information in the GIS provides a convenient way to add a spatial component to the management of substation drawings and documents either directly or in concert with a commercial document management system.

Managing Transmission System Development

Transmission operators perform system simulation all throughout the day to help understand what will happen in the next hour or even minute. To do that successfully, they need an accurate model of the transmission system. The GIS provides the data about connectivity, materials, and of course location. Transmission system planners need the same information about the network for future planning and development. They need to know what plans have been approved and when. So, if planners are studying the needs of the network in five years, they need to know what the network will look like based on the projects that they will complete between the present and the study year.

While not commonly done, the GIS can be a place to hold each future network configuration based on the projects that the company plans to complete. The GIS can identify what elements of the system they will add, modify, and remove so that the planners can have a projected complete transmission network representation at any point in the future. They can do this by tagging the equipment in-service date as an attribute or by creating different versions for different future years. Thus, the GIS can serve as a record of the progress of the transmission system from today forward to answer the question of what the system will look like in one year, two years, or five years.

Load Forecasting

The most basic input to a system planner's work is a projection of what the load will be in a future study year. GIS is a common tool for load forecasting for both transmission and distribution. The distribution planners of course will need more of a subregional approach, whereas the transmission system planners need a more macro approach. The workflows are similar and involve GIS spatial analysis.

The planners begin the process with a regional analysis of the current loads. From the consumption and loading information, planners create density maps of population. Planners look at the demographics and income levels for today's load mix. Figure 3.10 shows a simple GIS web map illustrating regions (polygons) of income levels.

Planners can use other data sources like the average age of the population. Demographic data such as population profiles (urban, green aware, NASCAR aficionados, people most likely to buy electric cars, and so on) is well known and easily

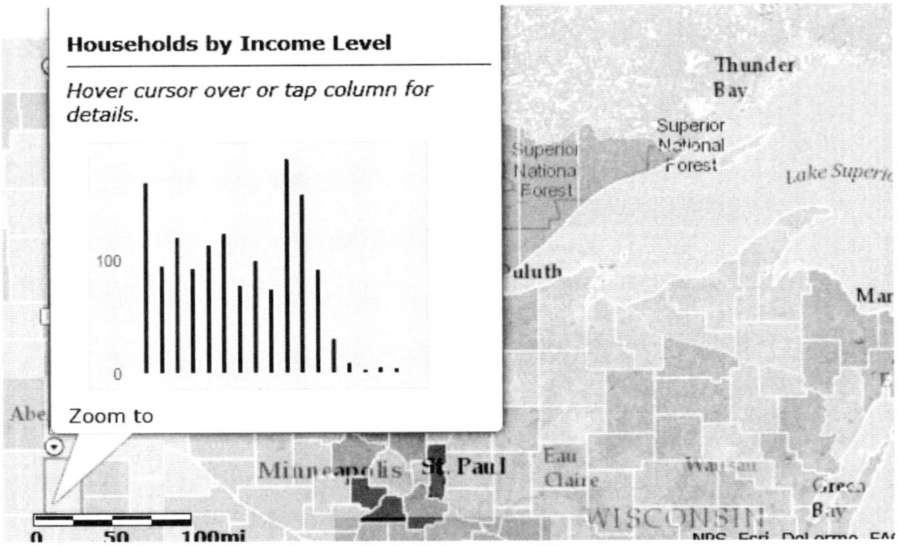

Figure 3.10 GIS web map showing income levels. (Source: Esri at al.)

obtainable. Each of these data sources is displayed as a map layer. The spatial analysis can associate load levels with a weighted overlay of these different data sources to get a picture of how the current electric load patterns are associated with different demographics. Other nondemographic data can be added to the analysis, such as land that is not developable (e.g., protected wetlands, mountainous regions, and badlands). The analysis can include present and projected zoning information from community master plans.

Once the planners complete the spatial analysis for the present day, they can then project these factors into the future using available prediction patterns from a number of different commercial sources. Factors such as planned transportation additions (like high-speed rail or subway systems) can be included. The net result of the load-forecasting model is a load density map. The final step is to overlay the current transmission system and perform a new power flow analysis based on the new load projections and see where deficiencies emerge.

In the past, planners simply took the load data (at, say, each of the HV/MV substations) and projected load growth uniformly. However, this ignores the significant variations in the regions. Figure 3.11 illustrates a simplified GIS load-forecasting analysis.

Line Siting

One of the most interesting aspects of transmission planning is the science of line siting. The concept was introduced in the introduction and this as noted was one of the earliest applications of GIS spatial analysis. Once the planning for a new transmission line is completed or once a demand for a new transmission line is established (say, for a new wind farm or power plant), the process of line siting begins. The process starts with: find the best route for a new transmission line from point a to b or from point a to the existing transmission grid. The best route is the one that minimizes the cost, the risk, and problems of permitting.

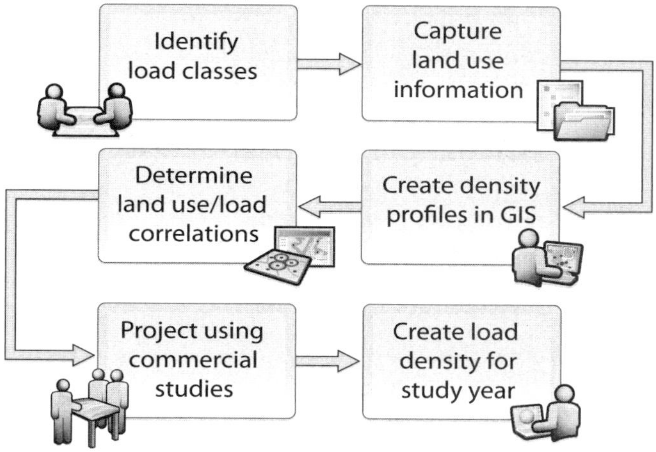

Figure 3.11 GIS based load-forecasting analysis.

This process is similar to the risk analysis. In the case of risk analysis, the line is already built. In line siting, planners want to limit the risk, so the factors are similar. They need to avoid areas of high natural risk, like fault lines, areas where tornadoes are likely to strike, and regions where it will be difficult to access the new line. In the line siting analysis, additional factors such as cost, the number of proposed bends in the transmission line, the construction difficulty (like crossing a river or significant change in elevation), sensitive environmental areas, aesthetics, avoiding population areas, and land availability are added to the mix.

The analysis follows the same pattern as other spatial analysis processes. The planners assemble all the input data sources. The sources are normalized to a similar scale. Weighting is added to each source. The result is a new map that illustrates areas of optimal route. The map combines all the factors together and displays the results as a graded or shaded map, and the darker the shade is, the highest the favorability is for the line. It doesn't show one solution but a map with continuous regions of favorability. It is not uncommon to do a preliminary design to see if the actual construction costs of the various routes are optimized. The routes then are sent to standard transmission design software packages, where costs are created. Then, the costs can be used as another input factor in the spatial analysis and the analysis reperformed.

The process is iterative. Once the planners create the preliminary analysis, they will often meet with community groups to determine exactly what the community reaction is. This meeting may result in adjusting some of the factors or changing the weighting. Permitting is often the most difficult and time consuming part of line siting, so having a GIS model that can immediately adjust factors provides a way to move the permitting process along much quicker.

Construction Management

The process of building a new transmission line or substation involves complex coordination of resources. Regardless of the project, whether it is a roadway, an airport, a substation, or a transmission line, nearly every aspect of the project involves location. The GIS can provide the project with the means to coordinate activities, to show progress, to show delays, and to help with the overall effort.

Managing Transmission Support Services

Whether the transmission operator is in the middle of a major upgrade to the facilities, building a new line, performing maintenance, or tracking down a fault, the transmission operator needs a number of support functions.

In order to stay out of trouble, the transmission operator must have an active compliance tracking mechanism. In the U.S., NERC has issued a number of regulations called critical infrastructure protection (CIP) requirements. While the prime driver behind these regulations is cyber security, the tracking of compliance can be facilitated by GIS elucidating where a company is compliant and where it is at risk. Of course, physical security is a high priority for a company with strategic and dangerous assets. The GIS provides visualization of the threats and helps to associate the threats with other activity in the area.

Supply Chain

Whether building or maintaining, utilities need to manage materials and people. They have to deal with spare parts, sourcing, and locating spare parts. When a transmission line is out of service, either during an unplanned or planned outage event, even if there is no customer outage, every transmission operator knows that the longer the line is out, the greater the risk of a catastrophic failure of the system. So, having ready access to the needed resources is essential. If a utility is investing in a new line, every hour of delay getting the line in service means lost revenue. So the support process of the supply chain is critical.

GIS can provide a means to optimize the locations where utilities can best stage material, locate the shortest path for routing of inspection crews, and identify areas of construction risk, such as the proximity to sensitive environmental or cultural areas. Chapter 8 will deal in greater detail with general utility supply chain issues and opportunities.

Corridor Management

Transmission operators have to access the land the transmission equipment is built upon. Often, the land is not owned directly by the transmission utility. The transmission right of way consists of a patchwork of parcels, with different owners and possibly different rights granted to the utility. The utility must understand the rights that the landowner has granted it. For example, a transmission operator regularly installs ground conductors with fiber-optic cables embedded in them. Or, the operator might install cell phone equipment at a substation. The utility may wish to lease a set of spare fibers to a commercial telecommunications provider. However, before they can legally do that, they have to understand if the easement or lease grants the utility the right to use the land for something other than the purpose of transmitting electricity. If not, then the utility cannot legally lease a fiber stand for commercial use. The utility needs to capture each parcel along the right of way and capture exactly what rights the easement in that parcel grants. This information is most often captured in or accessed by the GIS.

As stated earlier, the elevations of all the land along the right of way needs to be captured for clearances. During inspections, the utility has to capture encroachments along the right of way. Encroachments consist of any kind of material or construction that is not allowed in the easement according to the easement agreement. Common examples of encroachments are abandoned vehicles, sheds and out buildings, and paving. Encroachments can be unsafe and an additional risk factor for the transmission line segment. Encroachments are most often captured during the inspection process, which is performed using a mobile GIS application. As noted earlier, LiDAR is a convenient method of capturing encroachments.

Often a transmission line right of way is not located directly near paved streets and roads. It can be a challenge for the utility troubleshooter to gain access to the transmission right of way. The utility has no right to drive a vehicle across right of way abutter's land, even if the troubleshooter sees a problem. The only way a utility vehicle can legally gain access to the right of way is through an access road, which is either owned by the utility or the utility has gained an easement from the landowners. Since these access roads are rarely used, they can deteriorate or get overgrown with vegetation. It is important for the utility to clearly document the location of the access roads in their transmission GIS. Regular imagery captured along the right of way managed by the GIS is an excellent way to see the actual conditions of the access road.

Change Detection

One interesting GIS spatial analysis tool that transmission operators can use to help in corridor management is called change detection. Change detection is simply the comparison of two raster input files that were created at different times. Each pixel is compared producing an output product that highlights areas of change. This is helpful to uncover recent encroachments and to assess the condition of the access roads. If, for example, there is a bridge as part of an access road, change detection could uncover a washed out bridge. Change detection is especially useful before and after severe weather events and where flooding has occurred.

The Transmission Information Model

As noted in Chapter 2, an information model consists of a data model or data models, information products, and workflows. Perhaps the most important mission of the transmission operator is to keep the power flowing over the system. The second is to be absolutely certain that the company is complying with all regulations. So the operator needs information products that support this mission. In the long term, the operator needs to understand exactly when the system needs to be upgraded or added to. They also have to be keenly aware of the generation market, since in addition to keeping the power flowing, they are required to provide open access to any and all generation companies.

The Transmission Data Model

The transmission data model consists of data about the following:

- *The land corridor*—this includes all the ownership, rights, permits, access points, vegetation types, and conditions;
- *Surroundings*—the sensitive areas, habitat, hazards, historic weather patterns, and environment;
- *Real-time data*—the power flows, voltages, and outages;
- *Assets*—the location of every piece and part of the system or a way to locate all the assets;
- *The community*—who they are, what they are thinking, and the density;
- *The customers*—what their demands are;
- *The generators*—what their plans are and what their generation capability will be.

The GIS should not store all this information in a giant database. Instead, the GIS needs access to this information from a variety of sources, some of it created in other systems, some coming from internal systems such as SCADA or EAM and other information accessed from various agencies and governments. The GIS provides the spatial representation, the analysis, the dissemination, and the visualization.

The data model for the transmission system is really the collection of various raster data sets or layers, plus a detailed representation of the actual facilities that make up the transmission network, including the transmission lines and the high-voltage substations and generator substations.

Information Products

The main information product are the maps that show what is happening now, where there are faults, and how best to get to those faults. Since lightning and vegetation are the main culprits of line damage, transmission operators need information products that illustrate where the hazards are likely to exist. From an engineering perspective, they need GIS maps that locate every asset and its condition. Other examples of important information products:

- Boundary maps—jurisdictions;
- Overview maps of the transmission interconnection spatially reference;
- Schematics;
- Property acquisition maps;
- Environmental maps showing sensitive areas in proximity to oil-filled equipment;
- Spill-mitigation plans;
- LiDAR maps for line clearances;
- Future loads, proposed and current projects;

- Risk profile maps;
- Lightning fault maps—real-time and predicted;
- Situational awareness dashboard for network operations center;
- Land ownership and right-of-way maps;
- Right-of-way maintenance maps;
- Vegetation management polygon maps—status and future prediction;
- Vacant land inventory along rights of way and other land holdings;
- Reliability and performance maps—grading each segment of the transmission system;
- Flood prediction maps.

Workflows

Workflows take the data, sometimes user input, and display what the user will see on their screen, tablet, or smart phone from the web, their hard drive, or the cloud. The workflows that create the products are driven by the business requirements. The applications and workflows tend to be organized into four categories: data management, analysis, field integration, and awareness.

Transmission operators are faced with an enormous array of data sources; some of the data is within their control and creation, but much of it is controlled by outside forces, like the weather or angry citizens opposed to a new line or substation. The GIS workflows help pull the data together. One example is a maintenance application that helps workers bring data together from prior inspections, field hazard information, sensitive habitat, land ownership, current weather, and more. The application then helps maintenance planners craft the best daily work plan to optimize the cost and effectiveness of the maintenance plan.

The application that determines the best route for a new transmission line combines data management and sophisticated analysis to answer the question of what the best route is for this line. Applications for the field provide data management and analysis, and allow the field workers to provide feedback from actual situations to the decision makers. Finally, applications provide the utility with the situation as it currently exists, which brings data management, analysis, and field intelligence to the decision makers so they can shorten outage time, prevent damage, and keep workers and the public safe.

Getting data from other systems provides a rich environment for decision making. Keeping good historical records based on spatial data keeps utilities out of trouble and provides transparency. Making sure that all parties know the exact configuration of the system is critical and often very difficult to achieve.

GIS Matters for Electric Transmission

Transmission is spatial by nature. Transmission is everywhere. The GIS provides a convenient means to see all aspects of the system as it exists today as well as how it could look tomorrow. Historically transmission has been represented in schematic form. The vast majority of SCADA systems are schematic. While this provides an easy way to visualize the relationship of lines to each other and lines to substations, it does not provide the spatial context of exactly what's going on in the system.

Transmission operators are gaining a better appreciation of the value of spatial analysis for risk profiling, siting, and planning. GIS doesn't replace the design systems, the security systems, or SCADA. Instead, it compliments all of these systems by answering the questions of where things are, what's going on where, and perhaps even why that might be the case.

CHAPTER 4

The GIS Electric Distribution Facilities Model

This chapter will cover the distribution system itself, detailing the various components of a GIS facilities model. Chapter 5 will cover electric distribution business problems, such as new designs, inspections, asset and risk management, maintenance, and restoration. This is a growing area within the utility GIS users. Rather than restrict the GIS to technical information that they know, such as the location of a pole, the business applications are more about discovery or data enrichment. It combines the authoritative data (from the facility model) with business intelligence, such as predictive and real-time information, to gain business insight that leads to more informed decisions. Chapter 6 will address restoration and emergency management of both transmission and distribution. See Figure 4.1 for a breakdown of the workflows for the two different categories of GIS in electric distribution.

Distribution Business Versus Retail

In the era of vertically integrated utilities, the notion of retail business didn't exist. The distribution business handled the metering, billing, and customer engagement

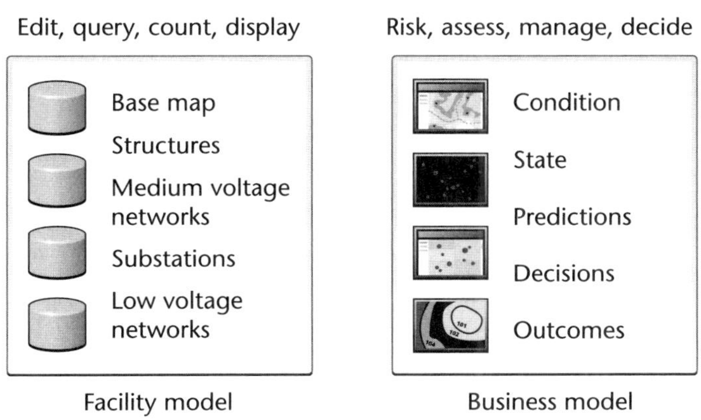

Figure 4.1 The application of GIS to the electric distribution business.

for the utility. Today, the tasks of metering, billing, and customer engagement define the retail business. In many countries and in several states within the U.S., governments have required formerly integrated utilities to unbundle their retail business. Further, the governments do not regulate the retail businesses, and therefore they can set their own prices and terms. However, the only aspect of the price that is not regulated is the price of the electricity itself, not the transmission and distribution delivery charges. This chapter and Chapter 5 will cover the distribution delivery business. Chapter 7 will address customer care, which for many distribution utilities is essentially the retail business. Many deregulated electric retailers or competitive suppliers can bundle other services such as broadband, gas, and telephone. As detailed in Chapter 7, GIS will play a strong role in the retail business, particularly regarding marketing new products, services, and demand response programs.

Who Owns the Meter?

The role of the meter in the delineation of the distribution business from the retail business is still not fully settled. As utilities roll out smart meters, the data from these devices have two users, the distribution business and the retail business. The distribution business needs the data to render the bills for the delivery charges (which include transmission charges as well.) The retail business uses the data for billing, too, but also for marketing other services. Soon, utilities will use the data to manage demand. Globally, the dust has not settled on the issue of who actually should own, operate, and maintain the electric meter—the distribution utility, the competitive supplier (the retail business), a third party who manages just the meters, or the customer. For the purposes of this chapter, the meter is part of the customer care organization, but the data about the location of the meter is important for both the distribution business and the retail business. Regardless of who owns it, the meter location or at least the location of the end point of the service line is managed by the distribution business and is normally managed by the GIS.

The Business Drivers for Distribution

The business drivers of electric distribution like the transmission business are as follows:

- Make money—that is, the company must make a profit to exist or, in the case of government utilities, not lose money;
- Keep customers happy—that is, to maintain high availability and quality of power;
- Keep employees and consumers safe;
- Be a responsible member of the community. It must be able to provide service to all that seek service.

To accomplish this mission, the electric distribution business must manage day-to-day operations, make sure that the infrastructure will support the future demands, provide connections to new customers, and provide support services to facilitate the mission, like having fleet support, logistics, and materials management.

Figure 4.2 shows the organization of a distribution electric business. The next two chapters will focus entirely on the distribution delivery tasks. Chapter 6 will detail emergency and operational management, related to power failures and restoration, and the following chapters will deal with customer care (the retail business) supply chain (the support for the business, finance and accounting, and corporate services).

To meet the lofty goals of making money, keeping customers happy and employees safe, and being a good neighbor, utilities recognized early on that they needed the right information. Given the nature of a distribution system where their equipment is scattered throughout their service area, they knew they needed a way to keep track of all of their assets. So they used maps.

Distribution System Mapping Started a Century Ago

While integrated electric utilities' use of GIS in the generation and transmission business has been spotty, they have modeled their electric distribution systems in some form of digital mapping systems since the 1970s. The reason is simple. The electric distribution facilities are everywhere. The number of elements in a distribution system is an order of magnitude greater than elements in a transmission system.

Results Have Been Mixed

Over the years, utilities have viewed the extent, degree, and success of these early digital mapping systems with some skepticism. Even today, many distribution utilities view their digital mapping systems, often misnamed GIS, as a network documentation system. They misname the system since many of these digital mapping systems are simply replications of their early hand-drawn maps on a computer. A real GIS can manage and process, spatially adjust, and fuse a variety of spatial data sets, enrich the data with spatial intelligence, and publish the result over the web to decision makers in the company and customers outside the company. The most common (and sometimes the only) workflow of these digital mapping systems is updating the digital maps with as-built sketches of new construction and maintenance, then printing and distributing the maps for use by field crews and maybe distribution planners—essentially the same workflow (only faster) that utilities had been using a century ago.

For the most part, utilities implemented early digital mapping systems to automate the mapmaking process, not to help run the business in a more effective

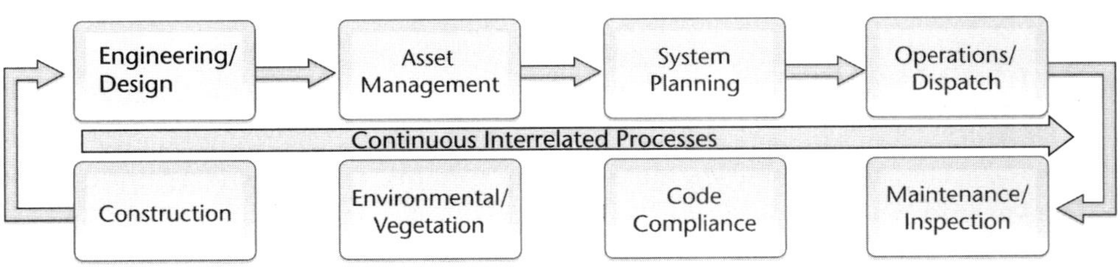

Figure 4.2 The common process of an electrical distribution business unit.

way. In fact, for some utilities a hard and fast requirement of the digital mapping system was that the output products had to exactly mirror the old paper maps. The only difference between the old paper maps and the computer-generated maps was that the text and symbols were clearer. Ironically, as it turned out, the early computer-generated maps actually had trouble replicating the old maps, because the drafting people could take all kinds of liberties with the placement of symbols and text. Since the use of paper maps and the old mapping grid systems dominated the workflow of utilities, digital mapping systems that simply replicated the old processes are not much more effective than the hand-drawn systems they replaced.

Why Paper Anyway?

The problem that early users of digital mapping systems were trying to solve was strictly a map production problem. The problems they were trying to fix were: how can they make changes faster (like a word processor for drawings), not have to erase large sections of the maps, and how to keep better records of exactly what is in the field? So the design of these early systems never foresaw the use of these systems for outage analysis, risk profiling, asset management, or as a critical component for smart grid.

The other issue is that since maps took a long time to make and the old hand-drawn maps were hard to edit, utilities put as much information into them as they possibly could. The maps then served all kinds of workflows, from restoration to engineering. With a modern GIS, the maps can be tailored specifically to the workflow. Despite all the advances in today's modern GIS, and technology in general, utilities still attempt to create their GIS maps to look just like the old hand-drawn maps.

A very simple example of a modern way of conveying information on a GIS map is in the use of pop-ups. Pop-ups are used extensively in smart phones and tablet apps. Text is kept to a minimum since displaying text takes additional computing power. Instead, most tablet mapping apps simply display atrtributes or informaiton about a feature or point of interest on demand. A user clicks on a symbol of say a point of interest, like a bank symbol and the system displays the details of the bank in a pop-up. In an old printed map, all of the attributes of every system needed to be printed. Figure 4.3 is an example of a simple electric distribution map with a pop-up that shows information about a device. Very little text is actually displayed all the time, unlike in a paper map.

The main reason that distribution digital mapping systems have had mixed results is because the use, development, and dissemination of the digital maps and ancient grid systems are tightly linked to old business processes and the comfort and familiarity users had with the old paper maps.

Digital Mapping Systems Evolve

Utilities and vendors coined the term automated mapping/facilities management (AM/FM) to describe digital mapping systems of electric distribution systems and gas and water distribution systems as well. Utilities adopted AM/FM systems primarily to document the location of electric distribution assets, create an editing workflow, and then plot the results. Once utilities created the plots, they would

Figure 4.3 Pop-ups help convey information on a map. (Source: Esri et al.)

distribute copies of the plots, just like in the old days, when they made copies of old linen and Mylar maps. Once the plotted copies arrived in the field, the staff treated them exactly like the old-fashioned maps. The new prints sat in a pile, waiting for someone to transfer the redlines from the old file copy print to the new one just issued by the GIS. Figure 4.4 is an example of how utilities marked up old paper maps.

As noted in Chapter 1, the main capabilities of a "real" GIS include data management, analysis, and awareness. Eliminating the idea that someone has to capture spatial information on a printed map or a map display that looks like a printed

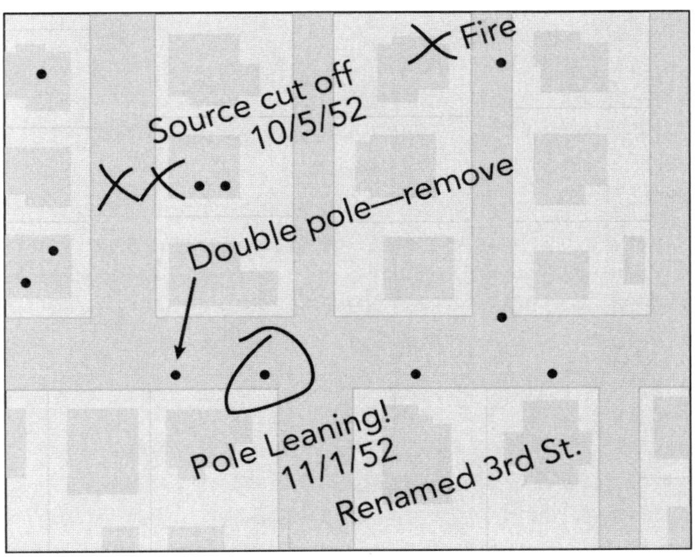

Figure 4.4 Typical redline markings on an old map.

map helps turn the conversation from the GIS being used as a better, faster map maker to using GIS as an information system. Field personnel can capture correction information in the field directly in the GIS, not by marking up a paper print with a red pen. Utilities can use the GIS to discover things that they can't see directly, such as the location of the electric distribution system that is most vulnerable to failure based on real-time information. Finally, the whole process of map production, editing, printing, and distribution can be eliminated entirely by providing direct access to the GIS to everyone in the company on the their desktop, over the Web, on their internet TV, or their smart phones. Figure 4.5 details the evolution of the electric distribution business from hand mapping on linen to cloud-based GIS.

The next several sections detail how the electric distribution system is modeled in GIS.

The Electric Distribution System Facility Model

In the early days of electric utilities, there were only two systems: generation and electricity delivery. Edison's plants generated the electricity and Edison's wires delivered the electricity to the homes and businesses. With the advent of the transmission system, the distribution system as a separate entity evolved. Today, the U.S.'s Federal Energy Regulatory Commission (FERC) defines transmission as any electric delivery system operating at 100,000V or greater. Similar definitions exist outside the U.S. That implies that any electric delivery system operating below that voltage level is something else: distribution. Old legacy transmission lines that operate at lower voltages such as 69 kV or even 35 kV don't really function in the classic sense as distribution lines. For the purposes of this book and this chapter, these subtransmission lines are just old transmission lines, and when utilities model them in the GIS, they treated them just like transmission.

The distribution network has two voltage levels: medium and low.

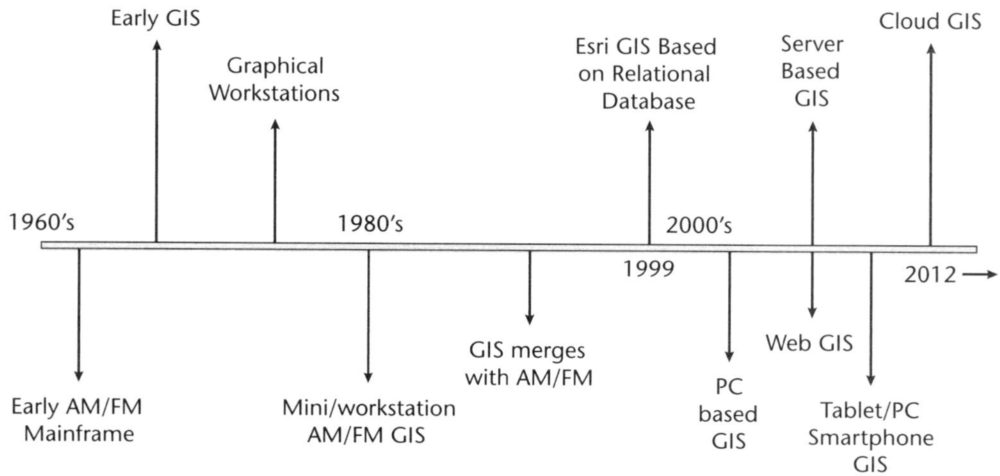

Figure 4.5 Timeline showing the progression of GIS and mapping at utilities.

Medium Voltages

Medium voltages have a wide range of values. The term medium simply means not high voltage, such as voltages common for transmission systems, such as 100,000V and above, and not low voltage, such as voltages below 1000V. Like transmission systems, the medium-voltage network has evolved over the years. Many electric distribution systems have several medium-voltage networks. This is not by design. Early medium-voltage networks consisted of equipment designed for relatively low-medium voltages, such as 2300V or 4000V. Over time, utilities installed equipment designed for higher medium voltages, such as 11,000V and 15,000 V. Utilities have adopted medium voltages at 35,000V for new installations. The reason utilities still operate lower-medium–voltage networks is simply that the older networks are adequate and the expense and disruption of a change out of these older networks does not make financial sense. To deal with multiple medium-voltage systems, utilities installed medium-voltage to medium-voltage substations. The substations can be as simple as a single three-phase step down transformer to a complete substation with breakers and bus bars. Even within a single nation, there may be a number of variations of medium-voltage levels, even within general categories. For example, utilities in the U.S. operate equipment rated at 15,000V at voltages such as 13,200V, 13,800V, 14,000V, and 14,400V due mainly to their past practice and the difficulty of reconfiguring their systems. This makes it difficult to interconnect medium-voltage systems from neighboring utilities.

Low Voltages

The low-voltage networks also can have several different voltage levels, but for more practical reasons. Normally, the voltage of the low-voltage network is the actual utilization voltage of the customers. In North America, the normal utilization voltage is 240V single phase. Most North American single-phase transformers have a mid-tap connection, which provides 120V for receptacles and small loads. Outside of North America, the low-voltage system provides 220V phase to ground based on a three-phase voltage of 380V. (This value also varies somewhat from country to country.) For larger commercial loads, higher low-voltage networks of 480V and even 600V three phase occur. In mesh secondary networks, voltages of 208V three phase are common in the U.S.

Parts of the Distribution System and the GIS Data Model

The facility model as captured in the GIS is the inventory of all the various parts and pieces of the distribution system located on a digital representation of the world, that is, a map, or base map. There will be considerable discussion of the base map or, as it is sometimes called, the landbase later in Chapter 8. Distribution systems have six parts, as illustrated in Figure 4.6.

The Electric Distribution System GIS Facility Information Model

These parts form the basis for the facility electric distribution system information model:

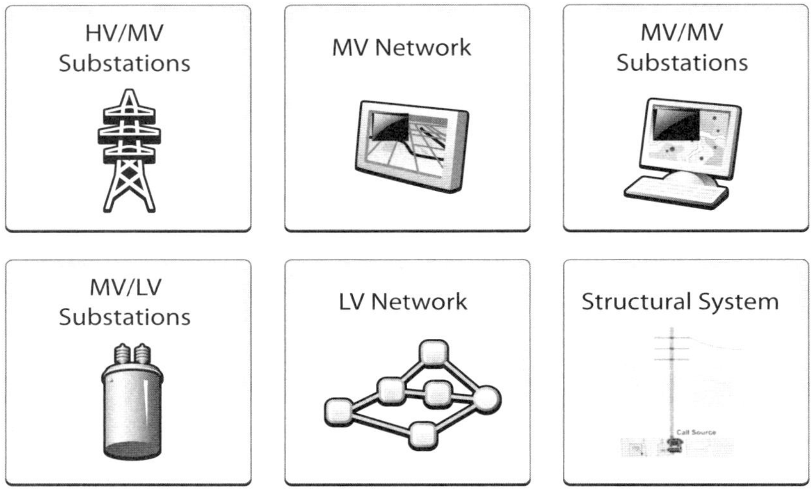

Figure 4.6 The key elements of the distribution system.

- *Distribution substations*—for the purposes of this book, these substations are called high-voltage to medium-voltage (HV/MV) substations.
- *The medium-voltage network*—these are the energized equipment operating at medium voltage, such as wires, arrestors, switches, capacitors, and voltage regulators.
- *Medium- to medium-voltage (MV/MV) substations*—these provide the bridge between newer medium voltage networks to older ones.
- *Low-voltage substations*—these provide the bridge from the medium-voltage network to the low-voltage network.
- *The low-voltage network*—these consist of the energized equipment that operates at low voltages, such as wires, fuses, and fault limiters.
- *The structural elements (noncurrent carrying)*—that support the distribution system, like poles, platforms, cross-arms, manholes, conduits, pipes, duct banks, pad mounted structures, trenches, tunnels and a host of other hardware.

The next several sections provide additional detail about the various parts of the electric distribution system, along with issues that often arise when modeling these components in the GIS. In addition, these sections include a discussion of the data model for each of the components at a high level. The next level of detail of the data model is the actual database table or schema for each of the parts or features. This book will not describe this level of detail. The detailed schema depends on the requirements of the GIS software selected, the industry standards that may be followed, the level and method of integration with other systems, the semantics used, and the information products expected from the GIS.

The core GIS attributes of the each of the various components are the unique identification of the feature itself and any relationship the feature has to some other feature. For example, the GIS needs to identify a regulating transformer uniquely by a name or identification tag. The GIS also must establish a relationship to its support structure location. The other key attribute of each feature is its connectivity to other current-carrying features, such as one conductor span to the next span. The GIS creates this type of modeling implicitly. The GIS has a built-in workflow that understands that if one end of the representation of a conductor is coincident with one end of another conductor, it creates the connectivity of the two conductors. Users can create a connectivity model in which the user explicitly creates connectivity by selecting which features are connected. Another approach is a hybrid in which most connectivity is implicit, but in special cases the user can override the implicit connectivity and force nonspatial coincidence connectivity.

A good rule is to only include attributes within the GIS if the GIS is the only system that creates and stores the attributes. Modelers often include many other attributes such as age of conductor, manufacturer, impedance, short circuit, and voltage ratings. However, other systems such as enterprise asset management (EAM) or meter data management (MDM) create and store the vast majority of these kinds of attributes. Just because users want these attributes to appear on an information product produced by the GIS does not mean that the GIS must directly store the attribute values. If two different systems update and store the same information, data synchronization can become difficult. The GIS can hold temporary data from other systems for performance reasons, but storing and editing of attribute data in one place is the best design.

HV/MV Substations

High-voltage to medium-voltage substations technically exist in both the transmission system (Chapter 3) and the distribution system, and provide the transition place between the two networks. Using a transport analogy, if the transmission is a wholesale bulk delivery system (such as cross-country rail) the substation would be the depot where material is off-loaded to local delivery trucks.

There are three sections of the substation: the high-voltage section, which consists of the tap off the transmission line, with switches and often high-voltage breakers; the transformer section; and the medium-voltage section. All the sections consist of three-phase equipment.

The High-Voltage Section

The high-voltage section will often have metering equipment that measures the power supplied from the transmission system to the distribution system. This becomes especially important when a different entity from the distribution utility owns the transmission system. In many parts of the world, the transmission owner and operator is a single entity within the country, often owned by the government. These transmission operators then supply power to distributors, some private and other public. The high-voltage section also includes equipment to protect and to isolate the HV/MV substation should problems occur within the substation or the transmission system.

The Transformer Section

The transformer section consists of the bulk power transformers themselves, any grounding equipment directly connected to the neutrals of the transformers, and breakers on the medium-voltage side of the transformer. Technically the demarcation point between transmission and distribution is buried deep within the power transformers. Since the HV/MV substation is more often owned completely by the distribution utility, utilities create the demarcation point between the transmission and distribution system at the tap point off the transmission lines.

The Medium-Voltage Section

Each transformer's medium-voltage winding supplies power to a medium-voltage bus, through a transformer medium-voltage breaker. A bus is a three-phase copper structure that facilitates the distribution of power to the various feeders. In cases where there is more than one medium-voltage bus, it is common design to connect each bus by a normally open bus tie breaker. Should a transformer fail or transmission line trip out, the bus tie breaker closes. This allows the feeders from both buses to continue to supply power to medium-voltage feeders. Medium-voltage feeder breakers connect to the bus. Each bus supplies a number of feeder breakers. The medium-voltage section distributes the power from the transformers to the various feeders that will distribute the power throughout the territory supplied by the HV/MV substations. Many GIS implementations do not include the HV/MV substations within the GIS and begin at the medium-voltage breakers.

Other Equipment in the Substations

In addition, substations have sophisticated control and relaying schemes that take automated action should a fault occur within the substation. They have backup power systems, automation systems, telecommunication equipment, annunciation, and a host of other equipment in the substation.

Issues of GIS with HV/MV Substations

Most early electric utility digital mapping systems focused almost entirely on the medium-voltage network. As noted earlier, the majority of electric utility GISs start or end at the HV/MV substation feeder breakers and include the low-voltage substations but don't always model the low-voltage network. That means that the HV/MV substations are regularly not included in the GIS at all. They should be.

Most electric utilities are organized along functional lines. The line groups manage the distribution networks. Often a different group manages substations. The demarcation point for line operations and substation operations is frequently the substation property boundary.

Utilities established mapping departments within the distribution areas, so it is not surprising that the maps and thus early digital maps did not include the feeding substations. Perhaps another reason for not including the HV/MV substations is that the personnel who dealt most with substations used schematic diagrams to visualize the operations of substations. Further, most SCADA systems had automated the control and monitoring of the HV/MV substations, so they had a tool that could serve as a way of analyzing these substations. The problem was and still is that without modeling the substations in GIS, even in schematic form, utilities find it difficult to do analysis on the complete distribution system. If an engineer needs to perform a relatively routine analysis to determine the cause of circulating currents through a substation, the engineer has to collect information from two very disparate systems, a SCADA system or a CAD representation of the substations and the GIS. Automating the analysis under these situations is difficult.

While utilities can represent substation equipment in their actual physical or geographical position, most utilities that model the substations in the GIS model substations in their schematic forms. Even in schematic form, the equipment can still maintain a reference to the actual geographic position, though not the exact location within the substation. For example, the GIS can perform a systemwide query asking questions like, where are there differential relays that the utility has not maintained in the last four years? To successfully accomplish this query, the GIS must have a link to information about the relays. As long as the GIS locates the busbar geographically and the link exists, the query will produce the results.

This is a good example of the GIS locating a piece of substation equipment, say, a transformer breaker, and linking it to other information in the substation, like relays. In this example, another system stores the relay information, such as a relay management system or work management system. Locating the transformer breaker spatially in effect locates all linked information such as relays spatially even though there is no direct spatial reference for the relays in the system that manages them.

The Data Model

The data model of the HV/MV substations includes the current-carrying features of the substation along with basic characteristics. The features in the GIS fall into several categories such as disconnecting devices, instrument transformers, monitoring points, conductors, transformers, and shunt devices. The data model can include other features, such as station batteries. Modelers could include other features as belonging to a building feature class of substations, which relate only to the building itself. This modeling belongs in the general category of facilities management. Utilities often include the tasks of facilities management in the supply chain support function. Table 4.1 summarizes the features of the HV/MV substation.

Depending on the information products (that is, the displays required by a particular workflow), attributes can be accessed from other corporate systems. For example, the date a substation mechanic repaired or inspected a transformer or the age of a transformer may be needed for a GIS query, but that data is better managed by an EAM system.

Table 4.1 HV/MV Substation Data Model

Feature Class	*Examples*	*Represented in the GIS as*
Disconnecting devices	High-voltage disconnect High-voltage breaker Transformer neutral disconnect Medium-voltage disconnect Medium-voltage transformer breaker Medium-voltage bus tie breaker Medium-voltage feeder breakers	Line or point, depending on the complexity of the model required. GIS software can allow point features to act like switches allowing current to flow or not. Export to network analysis modeling software often requires these point features to be modeled as a line.
Instrument transformer	Current transformer Voltage transformer	Point.
Transformer	HV/MV power transformer	Line or point, depending on the complexity of the model. Some models use a composite model with each winding being a point feature and the transformer itself is a container for multiple windings.
Shunt devices	Capacitor or capacitor bank Neutral devices Grounding transformers	Point. Technically, the device is a two-terminal device, but one of the terminals is ground.
Monitoring point	SCADA or DMS point	Point.
Conductor	HV bus work MV switchgear bus HV and MV connection cables Cables connecting transformers	Line.

Medium-Voltage Network and Feeders

The medium-voltage network includes the conductors, switches, fuses, and protective devices, such as lightning arrestors. They include voltage-control devices like capacitors and voltage regulators. The medium-voltage network consists of individual feeders that supply power to local neighborhoods, factories, and commercial facilities up and down the city streets and country roads mounted on overhead structures, directly buried in the ground, or encased in pipe, duct banks, conduits, and a host of other structures. Using the transportation analogy, the medium-voltage circuits are like the trucks that take material to from the depot to the stores.

The vast majority of customer outage occurs due to a failure or problem with the medium-voltage feeders. It is not surprising that distribution utilities focus heavily on the maintenance and upkeep of these feeders. Further, it's not surprising that most utilities started their digital mapping systems with the medium-voltage feeders since keeping good records of exactly what is in the field is critical for the good operations, maintenance, and construction of the distribution system.

GIS Modeling Issues on Medium-Voltage Feeders

The facility model of the GIS has the task of creating an easy-to-use and visualized view of the medium-voltage feeders displayed on a base map, usually along with the structures that hold or carry the medium-voltage feeders. It also has the task of representing an electrically correct connectivity model of the feeders. Utilities need to know what is connected to what for the purpose of tracing, fuse coordination, loading, determining transfer capability, planning, and providing the logical network to other digital systems like load flow and short circuit analysis programs, outage management systems, or distribution management systems. So this dual role of the model of the medium-voltage network (which is really the collection of medium-voltage feeders) of cartography and the basis for connectivity can be a challenge.

The original reason for the creation of the digital representation of the network was to improve upon the logistical mess of keeping feeder maps current and to deliver the most current maps to the various users throughout the company. So initially, cartography was entirely the focus of the digital systems. Later, utilities discovered that as they automated other parts of their systems, such as outage management systems, they realized that they needed network connectivity as well. However, by then, many utilities had already committed to a cartographic representation, so retrofitting these early systems became complicated. For example, from a cartographic perspective, a single line from one pole to another represents a three-phase overhead feeder, even though there are three phase conductors and a neutral. Additionally, a variety of insulator and cross arm configurations exists. If the utility wanted to show only the location of a single-phase MV/LV substation, they could represent that substation as a symbol at a point that connects to the medium-voltage three-phase line. Figure 4.7 shows an example of a simple representation of the single phase MV/LV substation. However, the single phase MV/LV substation in reality does not intersect the medium-voltage line at all. Rather, the actual construction consists of a tap conductor off one of the medium-voltage phase conductors that connects to one side of a fuse holder. Then, another single conductor connects to the medium-voltage bushing of the single phase transformer. Figure 4.8 shows this.

So the cartographic representation may not always represent the logical connectivity of the system. In this simple representation of Figure 4.7, the single-phase fuse is in the wrong place from a connectivity perspective. The transformer appears to be in series with all three phases of the medium-voltage feeder. This configuration is more like a regulating transformer. In fact, the transformer and fuse are actually tapped off only one phase of the medium-voltage feeder. Unless the modeler identifies the transformer as one that is tapped off phase *a* or *b* or *c* by including an attribute for the phase, the utility will have no way of knowing which phase the substation (in fact, a simple single phase transformer) is connected to. Even if the utility was diligent in making sure that they correctly updated the phase as an attribute, the utility could at some point swap phases, thus creating a situation where the topological model is incorrect.

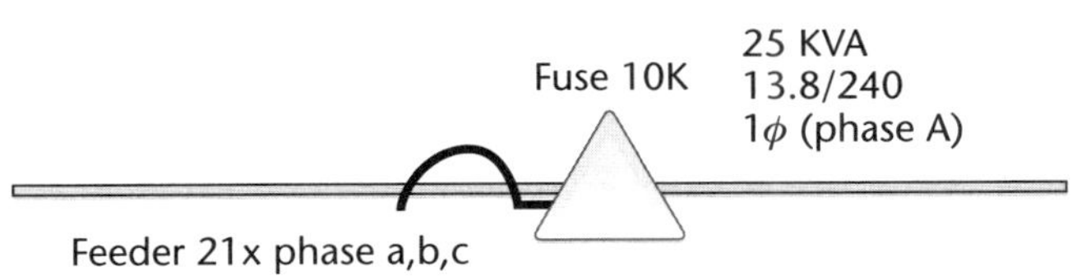

Figure 4.7 Simplified representation of MV/LV substation in GIS.

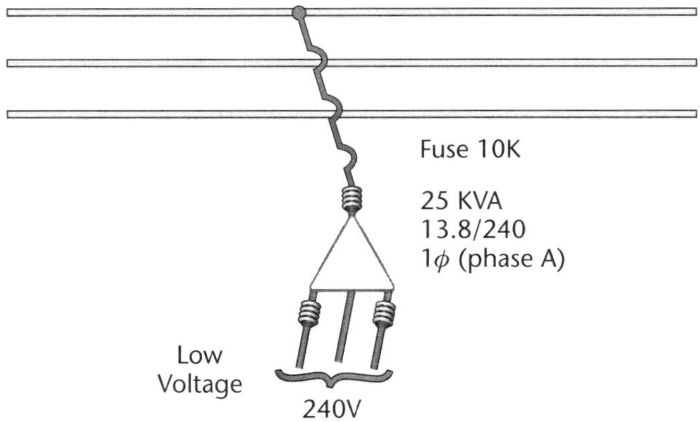

Figure 4.8 Actual representation of the MV/LV substation of Figure 4.7.

Esri performed a benchmark study called "Is Your GIS Smart Grid Ready?" at www.esri.com/smartgrid. In that study 250 utilities responded. One of the questions asked if the utility had accurate phase designation. Only 30 percent of the respondents thought that they adequately maintained phase information. The implications of this are that utilities won't be precise about the impact of outages, even with a complete smart metering implementation.

Utilities must balance simplicity of cartographic representation with the need for precise logical representation. Adding tiny tap lines clutters up the data and creates a more awkward representation. As utilities use GIS more and more as the foundation for their smart grid implementations, they will craft more flexible modeling options, such as scale dependency models that are simple at lower scales, showing a single line for all three phases zoomed out (smaller scales) and detailed three-phase representations at larger scales and perhaps detailed three-dimensional representations at even larger scales.

The representation of complex underground implementations poses significant cartographic and logical challenges. In many large cities, utilities built systems of underground duct banks containing a matrix of conduits. To represent each cable, let alone each phase of each cable, as a separate cartographic feature can be extraordinarily cumbersome. If, for example, within a 30-ft street in the old part of the city, the electric utility had installed 6 duct banks, each with 12 conduits, the total number of possible cables could be 72 three-phase cables. Even at very large scales (zoomed way in), the representation of 72 unique linear features within a 30-ft section of street is difficult. Assume that the scale of display is 1 in = 100 ft (a relatively large scale), a 30-ft street at that scale is less than one half an inch wide. So to display 72 parallel cables within less than half an inch would result in a meaningless display. Only until the zoom level at 1 in = 5 ft would the user be able to distinguish one cable feature from another. Then, if the utility wanted to display each phase and neutral as a separate feature, they would have to increase the number of cable features by four. Even at very large scales, they would be compromising the location of the cables.

So the simple underground GIS models will need to address the complexity of features within features for displays that conform to the specific situations, not a replication of standard map products based on traditional printed maps.

GIS Medium-Voltage Network Data Model

The data model of the medium-voltage network includes the current-carrying features of the network along with basic characteristics. The features in the GIS fall into several categories, such as disconnecting devices, in line transformers, monitoring points, conductors, transformers, and shunt devices. Table 4.2 summarizes the features of the medium-voltage network.

Medium- to Medium-Voltage (MV/MV) Substations

The medium- to medium-voltage substations connect one medium-voltage level to another. These substations can be as simple as a single three-phase pole mounted step down transformer to a complete substation with switchgear and control buildings surrounded by a chain link fence. The sole purpose of this type of substation is to provide a transition point between one medium-voltage network to an older legacy network. Like HV/MV substations, there are three sections: the high-voltage section or, more precisely, the higher of the medium-voltage section; the transformer section; and the lower-medium–voltage section.

Other kinds of substations include medium-voltage switchgear that distributes power to various parts of large campuses. These would not include transformers but would include feeder switches and medium-voltage switchgear.

The GIS could have significant gaps in the continuity of the medium-voltage network if utilities don't model these substations. Since many digital mapping systems and later GISs do not model substations at all, this gap is still quite common.

Table 4.2 Data Model of the Medium-Voltage Network

Feature Class	*Examples*	*Represented in the GIS as*
Disconnecting devices	Line switches	Point or line.
	Reclosers	
	Fuses	
	Disconnect switches	
	Isolating switches	
	Tie switches	
Transformer	Regulator or regulating transformer	Point or line.
Shunt devices	Capacitor or capacitor bank	Point.
	Lightning arrestor	
Monitoring point	SCADA or DMS point	Point.
Conductor	Overhead conductors	Line.
	Underground cables	

GIS Issues on MV/MV Substations

The representation of MV/MV substations is similar to the representation of HV/MV substations. The difference is there are two medium-voltage sections, not a high-voltage section and medium-voltage section. The higher of the medium-voltage sections is the primary side, while the lower of the medium-voltage sections is the secondary side. Each side will have disconnecting switches, breakers, cable connectors, and bus work. Transformers will step down from the higher voltage to the lower voltage.

What will change is the traceability of the phase conductors from one side to the other if the transformers have different phase configurations. For example, a transformer connected in a typical delta-wye configuration shifts the phase configuration. Phase 1 on the primary side does not correspond to a phase a on the secondary side. So if a utility wanted to know which customers are connected on phase 1 from the primary side of the substation, that query is impossible to determine on the secondary side (assuming a delta-wye transformer in between), since phase 1 may correspond to the line to line represented by both phases a and b. Modelers have to account for this. This situation exists even for simple step down transformers (which are simplified MV/MV substations).

GIS Data Model for MV/MV Substations

The data model of the MV/MV is quite similar to the data model or the HV/MV substations. Table 4.3 summarizes the features of the MV/MV substation.

Table 4.3 Data Model for the MV/MV Substations

Feature class	*Examples*	*Represented in the GIS as*
Disconnecting devices	Medium-voltage disconnect	Point or line.
	Medium-voltage breaker	
	Transformer neutral disconnect	
	Medium-voltage transformer breaker	
	Medium-voltage bus tie breaker	
	Medium-voltage feeder breakers	
Instrument transformer	Current transformer	Point.
	Voltage transformer	
Transformer	MV/MV power transformer	Point or line, depending on the complexity of the model.
Shunt devices	Capacitor or capacitor bank	Point. Technically, the device is a two-terminal device, but one of the terminals is ground.
	Neutral devices	
	Grounding transformers	
Monitoring point	SCADA or DMS point	Point.
Conductor	MV bus work	Line.
	MV switchgear bus	
	MV connection cables	
	Cables connecting transformers	

Low-Voltage Substations

Low-voltage substations bridge the gap between the medium-voltage level and the low-voltage level. They consist of the tap off the medium-voltage feeder, medium-voltage fused disconnects or switches, one or more transformers, and low-voltage switchgear. These substations can be single or three phase. The simplest example and a very common one of a low-voltage substation is a single phase tap off a single-phase medium-voltage line, no fuse, a single transformer, and no low-voltage switchgear. Most utilities in the U.S. do not refer to these implementations as substations. They commonly refer to this equipment as simply a distribution transformer. However, technically, this equipment serves the purpose of a substation.

Network and Spot Network Substations

A network substation is a special type of low-voltage substation. It feeds a low-voltage mesh network or a low-voltage spot network. These network substations are common in congested downtown areas where reliability is critical. Like other substations, they consist of three parts: the medium-voltage section with three-phase switches that isolate the substation from the medium-voltage feeder, the network transformer, and low-voltage section. The low-voltage section includes a specialized piece of equipment, called a network protector, which provides protection against the low-voltage mesh network from back feeding from low voltage to the medium voltage in the case of a fault on the medium-voltage feeder. These substations are often in sidewalk vaults or special rooms within office buildings or malls. The medium-voltage feeders supplying these substations are typically of underground construction and often are dedicated only to feeding these network substations. In fact, in most cases, even the HV/MV substations that provide the feeders to these low-voltage substations provide dedicated service to these very reliable network substations. Again, like other substations, utilities do not commonly model these network substations in GIS. Paper drawings or CAD files are often the only representation of these network substations.

Surface Mounted Substations

Common in suburbs and rural areas are MV/LV substations contained within cabinets mounted on fiberglass or concrete pads. Both three-phase and single-phase units are common. The substations are surface mounted, but both the medium- and low-voltage feeders are buried, sometimes in pipe, but regularly directly in the ground. Instead of medium-voltage switches, many of these substations have specialized connectors that can be removed with a tool that when pulled can separate the cable from the transformer medium-voltage terminal. These connectors are sometimes call elbows, because they resemble the shape of a human elbow.

GIS Issues for MV/LV Substations

Utilities model the medium-voltage connections for pad-mounted low-voltage substations in the GIS. Low-voltage substations range from very simple to very sophisticated network substations with all levels of complexity in between. As with other substations, the representation in a GIS is a compromise between a true spatially correct representation and a schematic one. For complex low-voltage connections, like in network substations, the GIS representation is hybrid with the overall representation of the equipment being spatially correct to a point and the connectivity of the parts, like the low-voltage fuses, buses, and switches represented schematically. Since most GISs represent the distribution system equipment on a two-dimensional map, vertical connections between the feeding lines and the substations are difficult to represent in two dimensions; thus, utilities take liberty with the actual physical location for the sake of logical connectivity. As long as the users understand the representation, this usually works well enough. This will likely change as GIS vendors add more and more three-dimensional capability directly in the GIS.

Low-Voltage GIS Data Model

The data model of the MV/LV substations is a simpler version of the other substation data models. Table 4.4 summarizes the features of the MV/LV substation.

The Low-Voltage Network

Low-voltage networks have several components:

- *Low-voltage mains.* Three-phase or single-phase wires or cables whose source is the low-voltage substation. These mains are generally radial in nature, but in some configurations they can be manually switched to be connected to other low-voltage substations at junction points. Some utilities call these low-voltage mains *secondary mains.*
- *Street lights.* Street lights are supplied from dedicated low-voltage circuits or are connected directly off low-voltage mains. The ownership of street lights is with the distribution utility, the municipality, or individual customers.

Table 4.4 Data Model of the MV/LV Substation

Feature Class	*Examples*	*Represented in the GIS as*
Disconnecting devices	Medium-voltage disconnecting switches and fuses	Point or line.
	Low-voltage fuses	
	Low-voltage circuit breakers	
	Network protector	
Transformer	MV/LV power transformer	Point or line, depending on the complexity of the model.
Monitoring point	SCADA or DMS point	Point.
Conductor	MV bus work	Line.
	LV bus	
	LV connection cables	
	Cables or wires connecting transformers	

- *Low-voltage services.* These are wires or cables that tap off the low-voltage circuits that feed directly to customer meters or service points. The ownership of the low-voltage services is with the distribution utility, the end use customer, or some combination of the two.
- *Customer meters.* The distribution system ends at the customer meter. Ownership of the customer meter is with the distribution utility, a retail competitive supplier, or rarely with the end use customer. The ownership of the meter varies around the world and is determined by the regulating authority.
- *Low-voltage generation.* While end use customers own the vast majority of low-voltage distributed generators, they still are part of the distribution system because they have direct impact on the low-voltage network. When they are generating more power than the customer is consuming, they supply power back into the low-voltage network.
- *Low-voltage mesh networks.* These networks consist of cables that are fed from secondary network or spot network substations. The cables are all interconnected forming a tight mesh of conductors. Should a fault occur in this network, the cable often burns clear with no customer outage.

GIS Issues with Low-Voltage Networks

The reason that utilities did not include the low-voltage networks in their GIS is probably because they are simple (like in North America). Another reason is that the cost of detailed conversion from the old paper maps of a lower priority system was too expensive. In any case, many utilities have not converted their low-voltage systems to any great detail. With the increasing popularity of local generation sources, like wind and solar systems, that connect directly into the low-voltage network and the increasing (or projected increase) use of electric vehicles connecting to the low-voltage network, low-voltage networks could be subject to increased stress. Thus, modeling this system will be critical so that utilities can determine if overloads are likely to occur.

Utilities have also neglected to model the low-voltage secondary or spot networks, which are extremely dense and complex. Very often, special groups within the utilities manage and operate these unique networks and have their own network documentation systems. Rarely do these mesh networks make it into their corporate GIS. One reason is that they rarely fail. However, when they do fail, the restoration can be quite difficult. In addition, if they fail, large sections of downtown cities will be without power. Another issue is that over time these networks can become compromised. In a conventional low-voltage network, failures always result in customer outage and the utilities simply repair what has failed. In a mesh network, failure of a cable does not result in customer outage, since there are so many paths the power can take to supply the customers. That's the good news. The bad news is that unless utilities have sensors and monitors along sections of the network, they never know when a failure has occurred. Documenting the network in the GIS allows the utilities the ability to model flows and uncover where gaps exist in the mesh as a result of equipment or cable failure.

GIS Low-Voltage Networks Data Model

The data model of the low-voltage network includes the current carrying features of the network along with basic characteristics. Table 4.5 summarizes the features of the low-voltage network.

Distribution Structural Elements

Structural elements of the distribution system consist of all the equipment supporting or encasing current-carrying equipment. There are several types of distribution system construction.

Overhead Construction

Overhead construction is the simplest to build and maintain. It is also the easier type of construction to represent in GIS. The most common structural element of the distribution system is the ubiquitous wood or concrete pole. Of course there are many variations on pole construction from cross arm, spacer, and wooden platforms mounted on poles. Most digital mapping systems and GISs represent poles, but few actually detail the variations in the construction. While wood poles are most common, Figure 4.9 shows an example of a concrete pole commonly used in the state of Parana, Brazil.

Wood pole construction is most vulnerable to severe weather, so it's not surprising that utilities focused heavily on mapping their overhead systems. Overhead construction consists mainly of wood, concrete, and steel poles with various types of mounting hardware. Utilities regularly capture pole information in the GIS. Figure 4.10 shows a typical wood pole.

Table 4.5 Data Model of the Low-Voltage Network

Feature Class	*Examples*	*Represented in the GIS as*
Disconnecting devices	Disconnect switches or points	Line or point.
	Isolating switches	
	Tie switches or devices	
	Mesh network current limiters	
Nonmetered loads	Street lights	Point.
	CATV power supplies	
	Telecommunication equipment	
Customer generation	Customer-owned generation	Point.
Conductor	Overhead conductors	Line.
	Underground cables	
Customer load points	Meters	Point.
	Customer connection points	

Figure 4.9 Concrete poles are very common in Brazil. (Photo by Bill Meehan.)

Surface System Construction

Surface systems consist of cabinets mounted on concrete or fiberglass pads. Utilities either directly bury the cables in trenches or pull the cables through conduit that they bury in the ground. Utilities capture the location of the pad-mounted equipment in their GIS. However, utilities often do not document exactly the location of the trenches or the conduits that make up the structural network. Figure 4.11 illustrates a simple cabinet that houses a single-phase surface- (pad-) mounted transformer.

Underground Construction

A complex network of underground manholes, vaults, duct banks, pipes, and conduits make up a sort of transportation network for the electrical cables. Utilities occasionally model the underground structural elements such as manholes, duct banks, and conduits but rarely in any level of detail. To gain information about these

Figure 4.10 Typical wood pole construction used for an overhead feeder. (Photo by Bill Meehan.)

structural elements, utility personnel have to refer to other sources of information like construction sketches or CAD drawings.

GIS Issues with Structural Elements

Utilities have a spotty record of maintaining the structural elements of the distribution system such as the poles, trenches, duct banks, conduits, pipes, manholes, hand-holes, vaults, tunnels, and service pipes in the GIS.

What is particularly important is the relationship of the structural elements to the current-carrying elements like the cables, conductors, and medium- and low-voltage equipment. During problems or when utilities need to construct new equipment on or within existing structural elements, workers lose precious time by trying to coordinate disparate sources of data. The main purpose of modeling the distribution GIS is to improve the awareness of the location, condition, and relationship of the entire distribution system, including the structural elements. As noted in Chapter 1, the GIS is able to perform spatial analysis, which can discover problems before they occur, but only if all the information is available. If, for example, a network vault is subject to tidewater flooding, then corrosion of equipment would be more severe than one that is not. Or, if that same network vaults need to be switched out, knowing that the vault is actually flooded when switching

Figure 4.11 Typical surface-mounted transformer cabinet. (Photo by Bill Meehan.)

is needed would be good to know, before the crew shows up and is faced with a situation they were not prepared for.

The Density of Underground Structural Systems Is a Problem

Underground structural systems proved to be challenging to represent cartographically. This is the same problem as noted earlier for the cables as well. If, for example, mappers wanted to properly show a 12 × 12 duct bank down the middle of a city street, they would somehow have to show 144 conduits in a section of a 30- or 40-ft street. At scales of greater than 400 feet to the inch, this would be impossible. Mappers had to create additional products like reference sheets, duct bank plans, manhole elevation drawings, pull sheets, and a variety of other drawings to be a able to understand the complex configuration of the underground structural system and the relationship of the structures to the electric networks that they house. For each new product they developed, they created a new problem of having to synchronize all the related data together. The utility might have created, printed, copied, and distributed a feeder map on day one, while the utility might have created, printed, copied, and distributed the associated manhole card on day 100. In the meantime, field crews and dispatchers don't have a consistent picture of the exact configuration of the electrical network and the structural support system at any given time.

GIS Data Model for Structural Systems

There are three unique types of structural systems: the simple overhead, surface, and underground network. The overhead system consists of discrete locations of poles. The surface system is a connected network of cabinets, trenches or conduits, splice boxes, and customer termination points. The underground network is complex system of rooms, vaults, manholes, splice boxes, and customer termination cabinets connected together by duct banks, containing conduits, tunnels, and pipes that form a complete underground network. Table 4.6 summarizes the structural systems.

The structural system by itself is a network similar to a subway system with tunnels and stations or a storm water system with catch basins and piping. If modeled as a network, utility planners can trace the network to discover proposed routing of new cables or to determine possible paths and distances through the network. Since the purpose of the strurctural network is to contain the current-carrying equipment, such as cables, switches, and transformers, the data model can include the relationships of the electric systems to the structural systems. For example, a section of cable is contained within a conduit that is contained within a duct bank that connects between two manholes. Another relationship is a surface-mounted transformer contained within a surface-mounted cabinet that is the termination of two conduits, which connect to other surface-mounted cabinets. It is common for GIS modelers to combine the cabinet with the contained equipment as a single feature, yet they are really quite different. A cabinet not only serves as a container for the transformer, but also as a slicing and pulling location.

Modelers also often combine underground cable with the containing conduit as a single feature, but they are different and from an asset accounting perspective are two unique assets with different attributes. For true asset management

Table 4.6 Data Model of the Structural Systems

Feature Class	*Examples*	*Represented in the GIS as*
Electrical rooms	Substation building	Polygon or volume.
	Vaults (building or street)	
	Manholes	
	Handholes	
	Low-voltage boxes	
	Pad-mounted cabinets	
	Customer connection box	
Linear structures	Tunnel	Line or volume for 3D representation.
	Concrete encased duct banks	
	Conduit	
	Service pipe	
	Trench	
	Riser pipe	
	Overhead route (the path of an overhead line)	
Point structures	Pole	Point.

and asset accounting, utilities should account structural systems separately. Rather than combining conduit and cable together (or ignoring the conduit in the model entirely), the data model should separately account for the structural element and build the spatial relationship between the structural elements and the electrical elements. Often those relationships are one to many and even many to many. Rarely do GISs keep track of exactly which conduit contains exactly what cable. However, this information can be critical during a fire in a manhole or a dig in to a duct bank.

Utilities have been more rigorous in the collection of pole information. Many utilities build the relationship between pole and electrical equipment. The reason for this is probably because poles are more vulnerable to damage.

Summary of the Current State of GIS and the Electric Facilities Model

This is the current state of the GIS facilities model for electric distribution:

- Many utitlity GIS implementations consist only of the complete medium-voltage feeders—precise phase designations are often missing.
- Structural elements are included where there is medium-voltage equipment.
- Rarely are underground cables associated with the structural elements that contain them.
- High-voltage to medium-voltage substations are often not managed in the GIS—the documentation consists of one-line station schematics and as-built design documents, not integrated with the GIS.
- Low-voltage systems substations (really medium-voltage to low-voltage or MV/LV substations) are modeled.
- Low-voltage networks (including low-voltage services) are not universally modeled.

The utility organization structure can sometimes result in fragmented data sources for the successful operations and planning for the distribution system. The implication is that utilities spend time on data synchronization and harmonization with no one system the single source of the distribution infrastructure data or at least the single source to be able to access distribution infrastructure data. Since the vision of smart grid relies heavily on information and particularly the distribution system, a complete electric distribution system GIS will be required.

Cartographic representation of the distribution can be a challenge due to the density of the equipment involved and difficulty in representing vertical information or essentially three-dimensional data in a two-dimensional form. For example, the middle phase of an overhead three-phase feeder is often right on top of the pole. The low-voltage circuit is nearly directly under it. So attempting to display both the medium-voltage feeder together with the low-voltage circuit or circuits in their exact locations is impossible in two dimensions, so modelers have to make some cartographic compromise, which usually involves offsetting the medium- or low-voltage conductors or accepting that they cannot display both medium and low-voltage conductors on the same display. More use of three-dimensional tools in the GIS will lessen the need for these cartographic compromises.

Additional Challenges of Distribution System GIS

If things were that easy, they would have done them long ago is a common lament. Early mappers and the GIS technicians that followed had to balance two very different demands on the map. The first is to make a linear network system cartographically pleasing. That meant that the cartographic representation of electric distribution lines had to be easy to understand. In the old days before computers, mappers created symbols for things like poles. They represented a pole as a circle and the mappers tried to locate the pole symbol as close to the same location as the pole was in reality. Then the mappers had to create symbols for the overhead wires. This represented a more challenging prospect, since most overhead medium-voltage lines consisted of three, sometimes four, wires: the three phases and the neutral. The problem was that the three overhead wires are separated apart by several feet and the neutral is actually below the three-phase conductors. At large scales, the mappers could show each wire in their approximate location. However at small scales (higher zoom levels), the wires would be too close together to be able to displayed correctly. At, say, a scale of 1 inch = 1,000 feet, a distance of 4 ft is 0.004 of an inch on a display (or drawing), so the distance between conductors is smaller than the width of a line on a drawing. Mappers decided to use a single line to represent all four conductors, since at small scales, actual location was impractical. This worked well for a single feeder on a pole, but for more than one feeder the mapping problem got more difficult. If the mappers wanted to add the low-voltage mains to the drawing and had multiple medium-voltage lines, the cartographic problem became difficult. To solve this problem, mappers created one set of drawings for medium-voltage at one scale and another set for detailed low-voltage maps (if they did this at all). As the network grew in complexity, the maps become more and more difficult to manage.

The Problem of Scale

Even though feeders had three and often four actual assets between poles, mappers created a one-to-many representation and often were not able to capture the exact location of one phase to the other. As mappers added symbols for equipment on poles or in manholes, they found that to capture the exact connectivity of the equipment, like a switch, lightning arrestor, or low-voltage substation, they had to comprise connectivity and asset correctness to be able to represent the equipment cartographically so viewers could understand them. If multiple medium-voltage equipment was on a single pole, mappers would have to offset many symbols from the pole line.

At the heart of the problem is that many utilities do not take advantage of the ability of the GIS to creatively model assets. They continue to use the GIS to replicate the same cartography of the old paper mapping systems. Fortunately, this is changing rapidly, perhaps by the popularity of tablets and smart phones. Users understand that not every display has to contain all the information one will ever need. It is the interactive nature of the web, tablets, and smart phones that effectively convey just the right information needed for the task at hand.

The Facility Information Model for Electric Distribution

As noted at the beginning of this chapter, the most common and most developed use of GIS at an electric utility was for the electric distribution, so at this stage in the history of utility GIS, the distribution system is the most mature (compared to generation and transmission). It is also the most complex in that it has the most number of elements. Updates to the distribution system GIS are done frequently due to the number of changes utilities make to the distribution network. Crews replace lines and cables, adding new services, fixing problems, and inspecting the system every minute of the day.

Facility Model Workflows

The workflows to simply keep the facility GIS current involve simple tasks often starting with identifying a feature, selecting it, and then doing something with it. That involves editing, adding, moving, rotating, deleting, changing the symbol, building a relationship of one feature to another, and editing an attribute. A workflows for finding equipment is to query by name, draw a box around an area to select, trace a feeder to find all the equipment connected, pan and zoom, and then count specific features by area, by query, by trace, and by simple selection. These tasks are all involved in the simple workflow of maintaining the facility model. Figure 4.12 illustrates a web map showing the details of the electric facility model.

Advanced workflows involve performing quality checks on the data, like making sure that sections of the feeders are not isolated, testing for connected loops in a radial system, or checking for missing relationships like a feeder not connected to a pole.

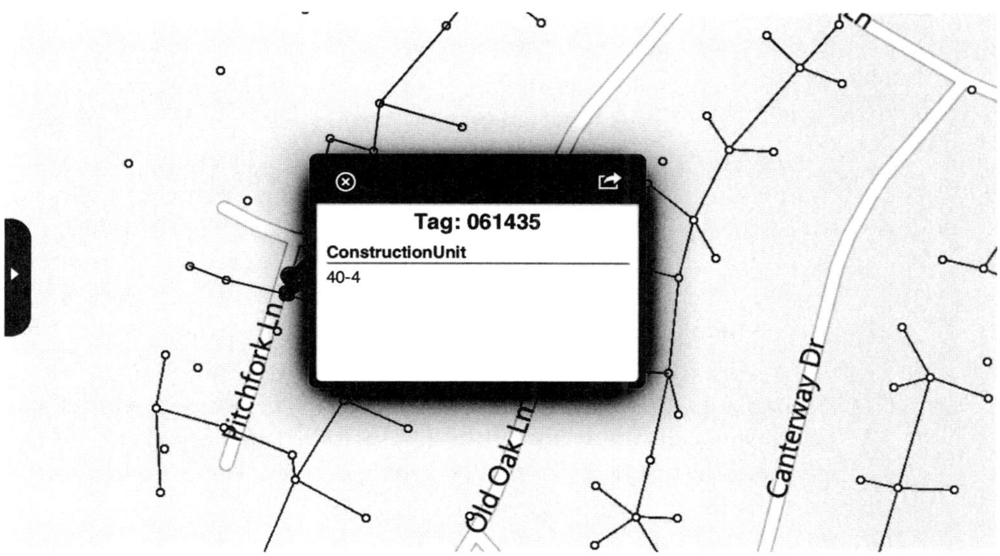

Figure 4.12 Web map showing the electric distribution facilities model. (Source: Union Power.)

Facility Model Information Products

The final element of the facility information model is the set of information products. These products are just the documentation of the various parts and pieces of the network as they currently exist. They do not include other applications that result in form of analysis. These business-oriented workflows, information products, and data models will be described in Chapter 5 and 6. Due to the complexity of the distribution system and the various levels of detail required by the utility, the information products can be extensive. Utilities will often define output templates that describe the scale of the output product (whether printed or displayed), the level of detail, the color scheme, the line weights and format, the extent, and any other information of interest. Normally, the utility will organize the output products by component, such as by substation, by feeder, and by voltage. The GIS functionality to create these products involve a query (i.e., find all the elements or features that meet a certain criteria), such as find all the medium-voltage conductors belonging to feeder ABC. It also adds the query, include all switches that are controlled by SCADA. The query could then add more information, such as to include only MV/LV substations that are three-phase. Finally, the query could be structured to create an output product at 1 in = 400 ft, display pole number for only those poles where medium-voltage equipment is located, and show normally open switches in red and closed switches in green.

Some examples of typical information products derived just from the facilities model are as follows:

- *Overview maps*—showing high-level medium-voltage feeders and substation location with only feeder tie switches shown. Often, the system displays each feeder in a different color. This is probably the smallest scale of all output products, zoomed out.
- *Medium-voltage feeder maps*—these show the layout of each feeder often by itself with all the medium-voltage equipment shown. These would be at a larger scale than the overview maps.
- *Medium-voltage feeder switching diagrams*—these are schematic representations of the feeders and are not to scale. These are derived products. They often only show main switches.
- *Low-voltage maps*—these show the low-voltage circuits at a very large scale since the level of detail is quite high. Often they are produced in a map book format by defined by standard grid boundaries. These are normally highly detailed, showing cadastral data, such as street easements, travelled way, building footprints, address ranges, street lights, and other small loads. These maps would detail service connections to customers.
- *Secondary network maps*—where secondary or mesh networks exist, these maps show all the detail of the low-voltage cable connections, along with all the cadastral detail of the low-voltage maps. These are essentially special cases of the low-voltage maps.
- *Station diagrams*—these are the schematic representations of the substations, and not built to scale. In the GIS, substations are displayed only in schematic

form, so these would simply be redisplay of the information contained in the GIS.

- *Conduit and duct bank layouts*—these represent a set of output products that detail the structural network of duct banks and conduit.
- *Manhole, building, and vault layouts*—these are layouts of the inside of the various electrical rooms that detail exactly the configuration of the conduit and duct banks as they enter and exit the buildings, manholes, and vaults.
- *Underground follow sheets*—these are geographically correct representations of the feeders as they traverse the underground conduit and duct bank system. These show the route of each of the feeders through the structural network.
- *Pole diagrams*—separate output products can be produced that show just the location of poles, or this information can be conveyed on the low-voltage circuit maps.

There are any number of information products that the GIS can produce for any combination of features, scales, and representations. Given that utilities are using mobile devices more and more, they will need to manage the amount of detail on a display carefully. In fact, utilities need to work at limiting the amount of detail and instead use interactive processes to produce the information they need at the time, like the use of pop-ups versus extensive labeling.

The Facilities Model of the Electric Distribution System

The distribution system consists of many parts. To be complete, utilities must account for and be able to locate each piece and part of the electrical distribution network. The focus of this chapter is on how to model the core components of the system. The next two chapters will focus on what additional functionality utilities can be add to the GIS to leverage this data for operations, infrastructure management, and emergency operations.

Where the Facility Model Is Going

The common use of the GIS facility model has been for network documentation. Utilities then built more and more applications on this model, adding more and data to it. Integration with other corporate systems often meant extracting data from one system, transforming the data, and then loading the data into the GIS. This of course meant that the data existed in multiple systems. That is not necessary anymore. The GIS can serve its facility data in many forms. One form may be as a simple map service in which the vast majority of the information is published in high-performing cached format for rapid display, yet detailed levels can be accessed only when needed.

Modern facility models will migrate to two-/three-dimensional hybrids. Users will see conventional maps at small scales, but when zoomed in to large scales, they see the actual configuration of the conduits or overhead wires. These configurations will be rule based and intelligent. Even though the electric distribution GIS facilities model has been around the longest, it will develop and evolve to meet the needs of the users, and it will be easier to use, faster, simpler to display, and yet much more focused and effective.

CHAPTER 5

Electric Distribution Development and Operations

The electric distribution GIS has two very different uses—one for carefully and completely managing all the assets owned by the distribution company as outlined in Chapter 4 and one that helps utilities make business decisions. This chapter will cover how GIS supports the business applications of electric distribution. The most difficult part of an electric distribution GIS is the actual building of the facilities model. In fact, it is often so difficult and so time consuming that utilities fail to take advantage of the GIS to use the facilities model for business applications other than to just make maps and keep the data up to date.

GIS Means Business for Electric Distribution

In addition to acting as a network documentation system that feeds data to other systems, such as outage management systems or distribution management systems, the GIS is a decision support system, enriched by all kinds of information, from the results of a business intelligence system to the mashing up of predictive information (e.g., the spatial analysis model on transmission risk in Chapter 3). It is one thing to know where a certain pole is; it is just as important to know if the pole will fall over during a wind storm.

GIS for the Business

GIS for the business helps answer questions, such as, where is the company spending too much money on maintenance compared to other areas? Or, where are their places that are more vulnerable to salt contamination? Or, where could the company violate environmental rules? The key element is location, just as it is in the facilities model, but it is the use of the assets within the context of the distribution business that is important. The information model for the business GIS can be dependent on the connectivity of the network and the detailed relationships of components to one another, but it is more dependent on the location of where things happen or what relationships exist to business results. Information products are not limited to displays or plots of the distribution assets at various scales and colors but would include hot spot maps that show areas of highest risk of failure or of vulnerability, or area maps that show where spending is highest or lowest. Here the notion of spatial analytics is more important than detailed network documentation. The network documentation serves as one of many data sources used in the analysis.

Figure 5.1 illustrates the GIS-related workflows for the two types of business activities: network documentation and business applications.

The Two Categories of Business Applications: System Development and Operations

There are two major business categories for the electric distribution business: system development (building the system) and operations and maintenance (running the system). Figure 5.2 details the differences between the workflows for development of the operations and maintenance (O&M) of the network.

System development deals with planning for future load growth, assessing where the infrastructure needs reinforcing, and engineering solutions for system expansion to meet growth or to improve reliability. Using the GIS for system development requires designing, and building those solutions. In addition, this activity involves meeting the needs of new customers, who require the utility to connect them to the electric distribution network. The process of connecting new customers

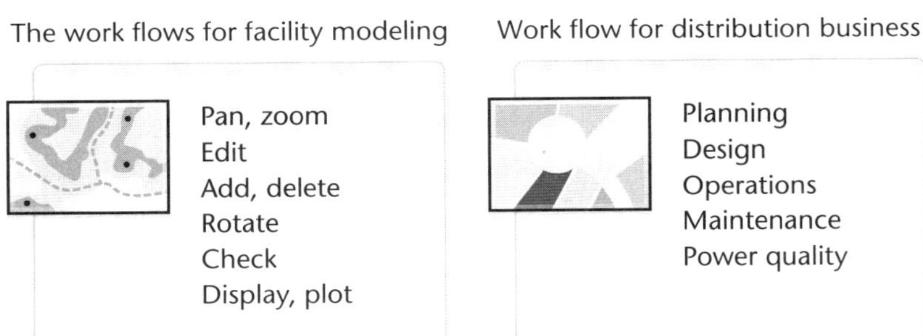

Figure 5.1 Workflows for each of the types of electric distribution GIS.

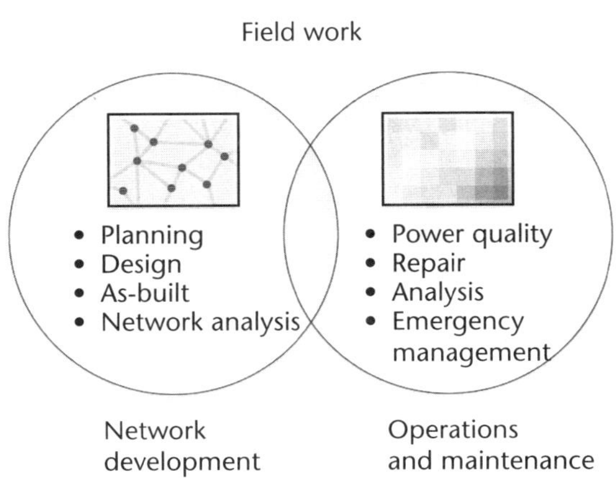

Figure 5.2 Business areas of system development and O&M and GIS impact.

is a customer care issue as well as an electric distribution business issue, and Chapter 7 will address this from the customer perspective.

Operations deals with the day-to-day activities of maintaining the system. It involves assessing the condition and making decisions about whether to replace or repair part of the system. If the decision is to replace part of the system, the replacement workflows are part of system development. The operations staff must respond to requests such as providing information to developers, dealing with blown street lights, and a number of other tasks. They schedule some of the tasks and others are ad hoc. The last section of this chapter will deal with these routine and nonemergency situations.

The most important activity of operations, however, is keeping the power on at all times. When the power does goes out, the operations people must put it back on. Chapter 6 will deal with outage and emergency management

Electric System Development Using GIS

Smart grid will change the complexion of the distribution network, but utilities will incorporate the smart grid initiatives alongside the routine development of the network to meet future loads and customer requirements. This section details the workflows of the development of the electric distribution system and the role of GIS. See Appendix B for a discussion of smart grid and GIS.

The Distribution System Never Stays the Same

The electric distribution system is in a constant state of change. That is because the distribution system links directly to end use customers. Customers move around, neighborhoods expand and contract, manufacturing plants open and close, and population shifts from cities to the suburbs and back again. Since the electric distribution system must meet the demands of the customer base, the utilities must perform a continuous state of assessment as to the adequacy of the distribution network, from the HV/MV substations to the low-voltage circuits to meet current and future demands.

Development of the electric distribution system has these major work processes:

- Distribution planning;
- Engineering, design, and construction;
- Documenting what has been built.

Distribution Planning

Utilities must perform planning on a regular basis and must look out into the future to predict what the loads will be one, two, even five years from today. Once they figure out the demands, they then must perform analysis on the electric distribution system to assess where the network is deficient and create a plan and projects to rectify the deficits. Of course, the planners have to account for and base their designs on the network as it is today. Further, they must estimate projects that might materialize from situations they do not control.

Predicting the Future

The key workflow for distribution planning is load forecasting. Load forecasting is the art and science of predicting the demand at some time in the future. Utilities look at load growth based on average values over the entire service territory (e.g., the load for the company will grow 2 percent or shrink 1 percent). Utilities will project load growth on a feeder-by-feeder basis. In both cases, the areas are too large or distances too long to rely on average values of load growth.

The load forecasting workflow is a classic example of GIS spatial analysis. The spatial analysis relies on spatial data sets, such as the location of businesses and data about demographics, zoning overlays, wetland overlays, vacant land, crime, migration patterns, income, road construction, and proposed projects. Each one of these factors comes from different sources, but each are map based.

As in the prior examples of transmission risk and siting, taking different data sets, converting them to data sets based on a common projection, normalizing the scales of the data, then applying a weighting to each of the data sets forms the basis for the analysis. Figure 5.3 shows the result in GIS display showing areas of equal load growth.

Within an area of several hundred square miles, it would not be surprising to see significant variation of load growth. Planners would run this exercise for each year in the study period. Some data might have to be created manually. For example, if the government is planning to install a new off ramp from an existing freeway, planners might create a polygon of larger than normal load growth as soon as the contractors complete the off ramp, assuming that service stations, fast food restaurants, and other service business may relocate near the freeway and the off ramp. That manually created map is just simply an additional layer in the model. GIS statistical tools can convert point data to areas of equal growth intensity.

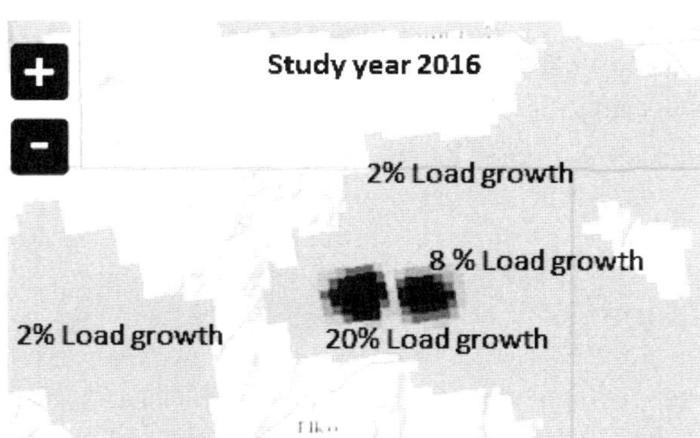

Figure 5.3 GIS map showing load forecasting polygons of equal load growth. (Source: Esri et al.)

GIS Can Bring Together Predictive Information from the Source

Many of the factors described earlier form the basis for the information products, which are a series of maps or map displays showing areas of equal predicted loads for each year. Using GIS, planners build GIS prediction models that access information from various sources, often using Web services, rather than having to extract, transfer, and load (ETL) and then store the data locally. They can, for example, consume a web service from a conservation commission that displays wetland boundaries and buffer zones or flood maps. The advantage is that once the planners build the GIS models, they can run the models regularly, thus tuning the forecast as things change. For example, if the freeway project is behind schedule, the modelers can move the project to the year the highway department completes the project.

So the planners create a load map for today's load profile, based on the information available, scaling it up or down based on the actual peak loads experienced. This provides a baseline for analysis.

The next step is to create a GIS map that displays critical electrical components and their current spare capacity. For example, a medium-voltage feeder section might have 10 percent spare capacity. Then the planners need to add to the facility map any additions and upgrades to the facility network that they know about for each year of the study period.

Finally, the model intersects the projected facility model for each study year with the load forecasting model for the same study year. The result is a map that highlights those key electrical components that will exceed their ratings. Contrast this with a simple load projection by area or by feeder. The GIS-based load forecasting model fine-tunes the analysis, by bringing in as much information as is known, not just by the utility, but by the entire community, and performs a scientific assessment of where demands will exceed capacity. This analysis will be even more critical as customers rely more on electricity for their transportation needs and as customers install onsite generation.

This analysis mixes very different data types, from predictive demographic studies that look at age, income, and buying habits to authoritative data such as zoning overlays to measured data like present load levels. GIS provides the platform to normalize this data in a systematic, repeatable, and scientific way. Figure 5.4 shows a GIS map that displays substations that will be overloaded in the study year.

Once the distribution planners determine that the company needs to build or upgrade a substation, feeder, or some other facility, they create a project to do so. The GIS plays a crucial role in that process, often called the work order process.

The Work Order Process

At some point, utilities receive a request for a change in electric facilities. Often that request comes from the distribution planners. However, there are many other requests that result in a project to develop the system. Someone needs a new service, a town needs some poles relocated due to a street widening, or a shopping mall is expanding, which requires increasing the capacity of their supplying transformer. As a result, someone at the utility has to create a design document that includes

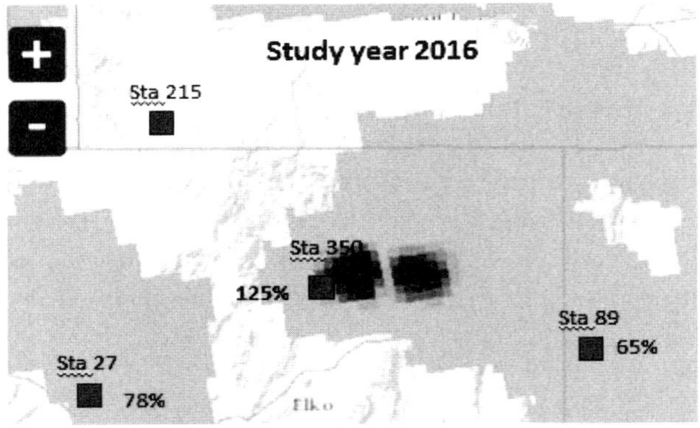

Figure 5.4 GIS map that illustrates when a substation will be overloaded. (Source: Esri et al.)

the detail of what the utility construction crews have to build, remove, or relocate. Often this document is in the form of a hand-drawn sketch or a diagram created on CAD system.

The GIS Design Process

Utilities recognized that involving the GIS prior to the construction of the facilities could simplify the entire workflow. If designers created the original design documents in the GIS, then they or someone else wouldn't have to re-create the design into the GIS after the crews completed the work. Since the people who actually created the designs were using map-based information to convey construction details to the construction crews anyway, using the GIS for the design documents was simple to understand.

Utilities get information from load forecasting studies that tell them exactly where the network needs to be reinforced due to the projected loading shortfall. As a result of these studies, they create projects, such as the following:

- Reconductoring lines—increasing the cable or conductor size to be able to carry the new load;
- Adding new transformers or low-voltage substations due to the increase in more localized load;
- Reconfiguring circuits to move load from heavier loaded feeders or circuits to lower loaded ones;
- Upgrading older lower-voltage medium-voltage feeders to higher-voltages feeders;
- Adding new breaker sections to substations, thus creating additional feeders;
- Adding new HV/MV transformers to substations;
- Adding new substations;
- Replacing legacy systems;
- Replacing obsolete equipment;
- Doing make ready work—that is, working on poles in preparation for new facilities, like smart grid cell relays or upgraded medium-voltage feeders;

- Creating automation projects—such as the addition of remotely controlled switches.

In addition to the planning process, there are other actions that trigger projects, unrelated to load growth. They are as follows:

- Customer requests for new service;
- Reliability of the part of the network falls below standards;
- Complaints from customers or regulators;
- Public works project that result in utility relocations;
- Major repairs due to emergency situations.

All of these actions create projects. The projects involve engineering, design, construction, and commissioning. Some projects are large (e.g., the building of a new HV/MV substation). Some are small, such as upgrading a single-phase transformer. All the projects have these things in common:

- Have a spatial dimension;
- Need detailed information as to what the current state of the electric distribution network is now;
- Need to create documentation as to what the state of the electric distribution will be;
- Need to create documentation to provide to the crews doing the construction—illustrating what they need to build;
- Need to create documentation of what the crews built compared to what the engineers and designers had specified to be built. This work product is called "as-built" documentation.

Since the GIS describes the facilities model of the electric distribution at a specific point in time, utilities have seen a need to develop a process that continues to keep the facilities model up to date. In other words, they needed a way to take the documentation that designers created and incorporate the new designs along with any deviations to the design that the crews documented back into the GIS. They also needed to make sure that it was kept up to date in an automated and controlled way, recognizing that the distribution system will not be static during the process of engineering, design, and construction. Dealing with an ever-changing network, utilities created a workflow called the long transaction.

The Long Transaction

The notion of the long transaction goes like this: the utility initiates a project resulting from any number of actions listed previously—a new customer connection, a reliability upgrade, or a public works project, for example. The distribution utility creates a work order for each of these projects. The utility receives an order to do some work (thus, the term work order). Often, the first thing the utility does is assign the project a work order number. From then on, they track all material, designs, costs, documents, invoices, and labor charges to this work order.

Early in the project, the designer needs to capture the current state of the electrical distribution network. In the old days before computers or GIS, the designer simply copied the operating map or maps of the area where the new work was going to take place. Then the designer used the map copies as the basis for the new design, documenting things such as which poles to remove, where to install a new switch, or where to install a new pole. Once the designer completed the work, the project then went on to others for approval, costing, assignment, and eventually construction. Sometimes the projects never actually were completed due to things such as a customer failed to get project funding, or higher priorities bumped the project, or perhaps a recession pushed off the need for additional capacity.

After the construction crews completed the work, they documented deviations from the original design. All of the documentation, from the original designs to the marked documents to material order sheets, were often placed in a folder called the work order folder. Eventually, the work order folders arrived at the desks of the people assigned to update the maps. The time from when the order was placed and the designer took the copy of the map to when the work order finally arrived at the mapper's desk could be days, weeks, or even months, depending on how big the project was or how slow and how many steps were in the process. The transaction is complete when the drafting people updated the old maps; plant accountants update the accounting data, reconcile all material, and close the work order. Unlike a short transaction, like the updating of a database field or correcting a simple error on a paper map, the transaction takes a long time—thus, the term long transaction. See Figure 5.5 for a graphic representation of the typical long transaction.

Note that the basis for the design was the original copy of the state of the electric distribution system captured at the very beginning of the transaction. However, since the network is always in a constant state of change, it is entirely possible that between the time that the designer took the copy of the network and the time the mapping person attempted to update the map, the network data could have changed. A car might have struck a pole and a repair crews replaced it with the taller one. A switch might have failed and been removed. Someone might have discovered that a piece of equipment was in the wrong place. So utilities needed a way of reconciling these changes in the area of interest of each design to alert the people working on the design that the design basis itself might have changed during this progress of the design.

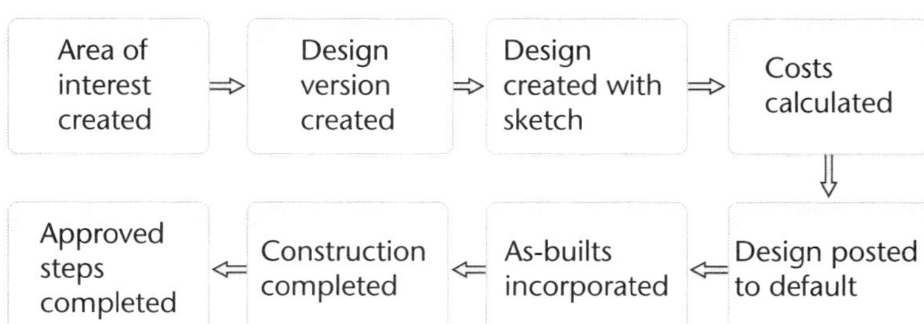

Figure 5.5 Typical steps of a long transaction in GIS.

GIS Design

To deal with this problem of long transactions, GIS implementations adopted the concept of versioning of long transactions. It works like this. The designers create the new work directly in the GIS itself. However, they can't simply update the GIS to reflect the new design, since this would eliminate the information about the present state of the network. Instead the GIS creates a version of the new design or even multiple versions of a new design, which are not disconnected from the as-is situation, but are sort of layers of what the network will look like when the crews complete the work. Utilities often name the as-is network data set the default version. The idea is that as soon as there are changes to the default version, the GIS can reconcile the outstanding versions with the changes in the default version. The GIS then notifies the in-process work orders there is a change in the design basis. See Figure 5.5 for a graphic representation of the typical long transaction.

Given that it takes perhaps several months for a work order to go from design to construction, it is entirely likely that something would have changed within the area of the work order during that time. The idea is to have the GIS constantly notify each outstanding design that something has happened within the area of the work order that may have an impact on the design. The implications of not having reconciliation is that by the time the crews got the designs, ordered material, and scheduled the work, the situation in the field may have changed from when the designer first copied the map. That means that the crews would have to either stop the project or have to adjust the work in the field based on some new and unforeseen situation. Figure 5.6 illustrates the concept of versioning within the GIS.

The final step in the GIS design is that the GIS mappers update the design version with the as-built changes and simply post the completed and corrected design version back into the default version. The utility then deletes or archives the design version.

The GIS Design and Construction Work Flow

The process of GIS design involves the following:

- Create a linked version of the area of interest of the design—a new version.
- Perform the new design work in the version.
- Perform all network analysis and calculations in the version.
- Continuously reconcile the design version with the default version.
- Resolve conflicts when there are discrepancies between the design version and the default version. For example, if an emergency crew replaced a 35-ft pole with the 40-ft pole due to a car accident, the design version would show a 35-ft pole, whereas the default version shows a 40-ft pole. The GIS design application provides tools to detect these conflicts and allows the users to resolve them as soon as a discrepancy is detected.
- Finally the design version is posted to the default version after all conflicts have been resolved.

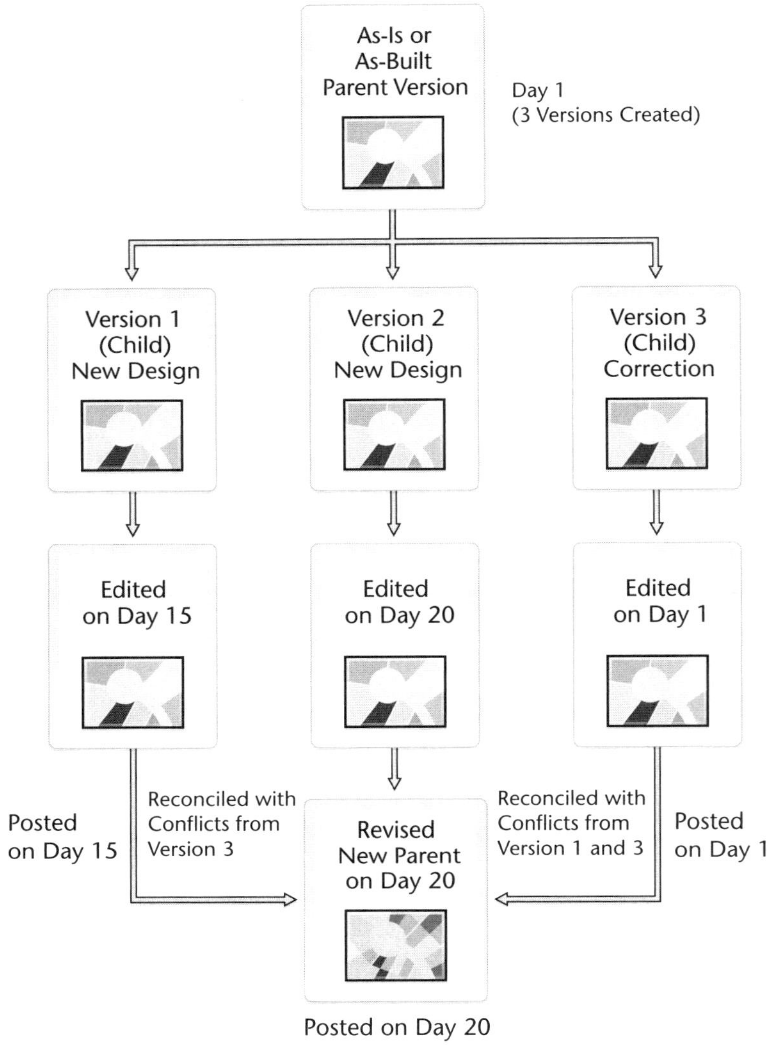

Figure 5.6 How GIS deals with versions for long transactions.

This process works extremely well for the work on the medium-voltage and low-voltage networks, and less so for additions and modifications to HV/MV substation work. Here, due to the complexity of the work involved and the time lines, utilities resort to conventional project management techniques.

Integration with Corporate Systems

Best practices for GIS design of distribution networks involve close integration with work management systems. During the engineering and design processes, utilities will often directly access work management systems or EAM systems to make sure that any new assets they add during a design is accounted for in both systems. Ideally of course, very little duplication of information exists in both systems. In addition, utilities will track and create work orders within the work management or

EAM system. So a linkage between the two is very helpful, if not essential. The GIS can display the extent of the work in a simple polygon in the default version, identified by a work order number. The EAM manages all other aspects of the work order, such as who initiated it, accounting information, descriptions, dates, and costs.

Using a GIS design shortens the overall design process, since the utility creates the original design only once, rather than having one person create the design and then another person re-enter the design features into the GIS after the crews finished their work.

Compatible Units

Some utilities have adopted the concept of compatible units to help with their distribution network process. A compatible unit consists of a list of all the material, labor costs, miscellaneous tasks such as police details, plant accounting information, and often a link to a construction standard. The idea is that designers do not have to specify the pieces and parts of design. Instead, they simply specify one or more compatible units at a point or between two places. If the utility uses compatible units for design, the GIS design version can show the location of each of the compatible units on the map, but the GIS can also access additional data about the compatible unit from the EAM or work management system through whatever integration framework is in place.

Longer-Term Projects

Utilities also need to document projects that they know they will design and build in the future. They do not need to detail the projects, only account for them at a high level for planning purposes. They can create these longer-term conceptual projects in GIS versions. This helps to do a better job of load forecasting. While the utility will not know the details of the design until later in the process, it helps to have the present and future information about the state of the network accessible in one place.

Integration with Network Analysis

During the design and longer-term distribution planning process, utilities must perform complex network calculations to actually determine the flow of power, short circuit issues, and voltage profiles. They also need to understand the extent of the losses that exist or will exist in the network. Utilities also need to know quickly when they receive a customer request for a new service, whether the network has the capacity to serve the customer directly or if they will need to invest in a major upgrade of some part of the system. If they need a larger cable or transformer to meet the proposed new load, they will need to notify the customer of any additional cost or delay that may introduce.

Much of the information needed to run a network analysis is contained within the GIS or is accessible by the GIS. So the workflow is for the GIS to export or extract that part of the network impacted by the new load. This may be simply the low-voltage substation and the low-voltage network, or it could be several medium-voltage feeders. The network analysis application extracts the network elements and their connectivity, extracts additional information, such as loads and impedances from other systems (smart meters, load estimates from the billing system), and calculates the power flow, short circuit, and voltage drops throughout the network. Often, the analysis modules present the results directly back into the GIS for visualization and dissemination.

Keeping the GIS Up to Date

The simplest workflow to accomplish updating of the GIS facility model is the as-built process. GIS users receive changes to the network based on the marked up maps, memos, emails, phone calls, or notes on scraps of paper showing changes or additions to the network. Those changes are a result of all the constant flow of projects that occur daily in the distribution network. This documentation indicates what the crews actually constructed in the field. As noted previously, collectively, utilities call this documentation as-built drawings or sketches. The users of the GIS simply update the facilities model using standard editing workflows, like pan, zoom, add, delete, move, rotate, and view, based on the information on these as-built documents. After they make the changes, utilities then publish, plot, or display the updated network.

In addition to simple editing, the GIS provides quality control within the editing process and often shortcuts to simplify the updating process. For example, the GIS would prevent an editor from connecting a medium-voltage cable feeder section directly to a low-voltage cable section.

Many utilities today use the GIS only for updating the facilities model. So the GIS in this case is simply a network documentation system (i.e., a system that documents where the location of the network elements are and provides some information about each of the elements, such as size or rating). The GIS used as network documentation is not to be trivialized. Many processes within the distribution business rely heavily on the information products that result from this process, such as those listed in the last chapter. Further, the utilities often extract the data from the network documentation system for input to outage management systems, distribution management systems, and plant accounting systems.

The facility model that the GIS manages, the as-built workflow that keeps it current, the workflows that extract data from the GIS, and the resultant information products are powerful tools for planning, design, and operations.

Issues with GIS Design

As noted the electric distribution system is in a constant state of change. Utilities add new customers continuously. They replace worn out equipment, extend lines, and replace poles struck by errant drives. Overhead electrical systems are constantly under attack by the elements. This means that it is a considerable challenge to have exact documentation of the electric distribution at any given time. The idea of the electric distribution GIS is that it represents the exact state of the assets in the field, their condition, and their connectivity. However, many utilities are unable to process the transactions in a timely manner. So there may exist a backlog of changes to the distribution system that the utility has not posted into the GIS. This may have been tolerated during the manual mapping days, but in the smart grid world, having unposted work in the GIS could seriously jeopardize the integrity of any smart grid operation.

Another issue occurs during the reconciliation process between one version and the default version or other versions that exist in the same geographic region. As noted earlier, conflicts occur when the existing state of the network changes while a work order or orders are still in process. What can happen is that these conflicts are not resolved quickly and can pile up over time, resulting in a bottleneck during the final posting process after the crews finish the work. The result is that the completed work orders are missing from the default GIS. Conflicts then create their own backlog. What some utilities insist upon is that when work order areas of interest overlap, the involved parties resolve the conflicts first before they begin the design. This involves a meeting. Sometimes the areas of interest can be adjusted so they don't overlap, and sometimes the parties just have to decide to communicate their designs to each other. The other option is to opt for feature locking of the design. Once designers specify an area of interest, the GIS prevents any feature in that area to be modified in any way. Then, if other designers attempt to create a version that encroaches on that someone else's area, they would get notified and would have to discuss their plans with the other designer. There are pros and cons to each situation.

False Conflicts

GIS modelers need to be careful not to artificially create conflicts by allowing GIS users to update attributes of features indirectly that have little impact on the design process. For example, an inspection process may update the date a pole has been inspected from one date to another. If, for example, the modeler created an attribute of the pole of date inspected, the attribute of the pole would have changed during the design process. During reconciliation, a conflict could occur that states that the pole has changed since the designer created the version. However, this conflict does not impact the designer's work and could hold up the work or cause needless extra work. So modelers must take care not to create conflicts that do not impact the design process. The key to conflict resolution is to prevent the possibility of conflicts in the first place by good communication among the parties engaged in updating the data.

Substation Planning, Design, and Construction

Once the distribution planners have completed the long-term load forecasting, distribution engineers assess the need for additional equipment. Sometimes the need is so great within an area that the solution is to build a new HV/MV substation or significantly add new equipment to an existing substation.

The process for siting a new substation involves finding the right blend of technical, environmental, social, and financial conditions. It's not enough to say that new substation needs be built in this or that location. Siting a complex structure like a substation involves community involvement, permitting, land acquisition, transmission extension, land suitability, impact analysis, and logistics. Most of the factors have a spatial dimension.

The general process is as follows:

- Gather the data.
- Perform spatial analysis on the data to provide a number of alternatives.
- Select the best several options.
- Present those options to interested parties in a way that is easy to understand.
- Prepare to adjust.

Utilities will need to gather all kinds of data, both spatial and nonspatial, from a variety of sources. Key data sources include the latest flood studies of the region, high crime areas, and environmental and aesthetic concerns. Since GIS can consume data from different sources and from different organizations, the ideal way to gather the data is to access the data at the source. For example, rather than obtain a collection of hard-copy maps from conservation commissions, municipalities, and police departments, it would be better to access the data through some kind of service, like a web service that is up to date, each time the data is accessed. Figure 5.7 illustrates the substation planning process using GIS.

Zoning and master plan digital maps from the government can help show the future landscape of any proposed area. The idea is to collect as much relevant data as possible and build a model that processes that data into a single map using the tools of GIS to discover the best locations for the substation, given all the variables and constraints.

Once the planners and the community have agreed to the substation location, the unity can develop the detailed engineering and design documents and then actually build the substation using the GIS to facilitate the logistics, the construction, the environmental monitoring, and the final commissioning. Once the utility

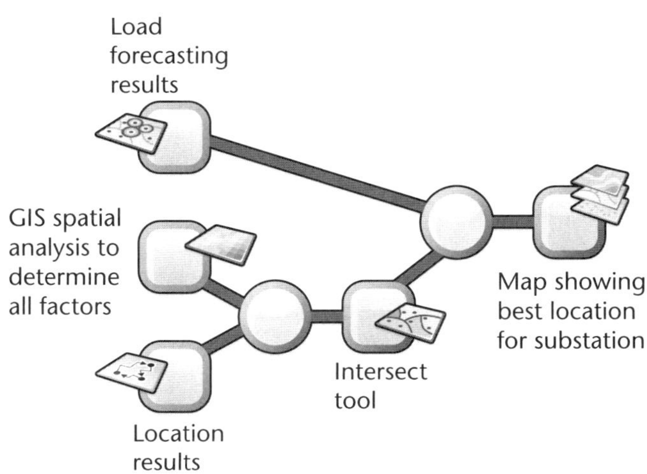

Figure 5.7 Substation planning process using GIS.

completes and energizes the substation, they can incorporate the one-line diagram into the GIS.

The Information Model of the Electric Distribution Infrastructure Development

Recall that the information model includes the workflows, the information products, and the data model. The workflows consist of the planning, the as-built process, and the GIS design process. The information products of the planning are the maps of the load studies, the areas of equal load growth, and the maps that detail where the distribution network will need reenforcement. The information products of the as-built process are those described in Chapter 4.

The information products for GIS design consists of the documentation of what the contractor or construction crews need to do to perform their work as the designer had specified. If the utility uses compatible units, the designers will include those units on the map display at their location. Utilities will include items like the bill of material (list of parts) for the work involved and often the cost of the work directly on the work order design.

Some utilities call these construction documents staking sheets. Utilities often perform the design in the field and during that design process, the designers would actually drive stakes in the ground to illustrate where they wanted the crews to install the equipment—thus, the name staking sheets. Other utilities call these documents work order designs or work order sketches. Regardless of the name, they are maps that show what work the electric construction crews need to perform to complete the work.

Today, for work in the field, mobile devices are becoming the tools of choice (versus paper maps). This allows the crews to use the mobile device to communicate changes to the designs. The workflow would be to deliver the designs to the crews to their mobile devices and for the crews to communicate any deviations to the office so that when the designers post the new design to the default version (or as-is version), they would easily see the deviations. The trend is that the communication from the field to the office is performed through wireless connectivity, sometimes using cloud computing as the broker to manage these services. Typically, the crews would indicate a correction or as-built change in the form of a redline change, not a direct correction to the design. Figure 5.8 shows a typical information product from a GIS design.

The data model for the infrastructure management process is an extension of the facilities model. Depending upon the architecture, the data can be included within the facility model or can be outside the data model and integrated with the facility model whenever needed.

Typically load forecasting data models consist of simply the services needed for the analysis, such as demographics, environmental overlays, and other information not typically under the control of the utility.

For the GIS design, modelers would design the system to access data from the EAM systems, such as compatible units. The facility data model would then be supplemented by the location of the compatible unit. However, the detail of the compatible unit would be accessed from the EAM, not stored directly in the GIS.

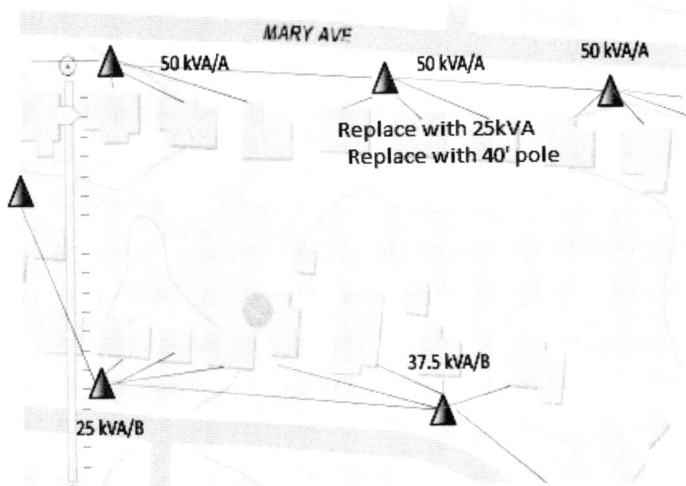

Figure 5.8 Typical GIS design. (Source: Esri at al.)

Summary of the Role of GIS in Electric Distribution Infrastructure Development

As noted in Chapter 4, the electric distribution consists of a large number of facilities, current-carrying and structural equipment, and of course customers. Utilities must know the location and relationships of this equipment to meet their mission. The better they know the state of their system, the more effective they will be. Historically, this has been a challenge for electric distribution companies. The process of keeping this data current can be daunting.

The workflow of designing a change to the distribution system to actually building it and tracking all the steps in between is the work order process. It involves starting with a solid understanding of the present state of the network as documented in the GIS, then continues with the documentation from the GIS of what will be altered (new transformers, poles, wires), communicating that information to many players in the company for approval, ordering material, and collecting payment. Then the utility must communicate those requirements created in the GIS to people who actually do the work. Those people can be utility crews or contractors. Once the work is completed, the crews document that the work has been accomplished and they indicate what if any deviations they may have made. They document those deviations in the form of as-built comments or symbols. The utility then updates the new work and any deviations from the design back into the GIS. The GIS then reflects the new work.

Requests for new additions can come from a variety of sources: the planning group, new customer requests, the asset management group, and outside parties like cities and towns who need the utility to relocate their facilities.

Utilities capitalize the work associated with the process. That means that regulators allow utilities to amortize all of the costs of performing any aspect of this work over a period of time. This is because the end result of the work is an increase in the overall capital assets of the utility. That would include the labor cost of a GIS technician incorporating as-built sketches into the GIS.

At any given time, the GIS represents the best source of what exactly exists in the electric distribution system if the utility is diligent in capturing all the assets in a timely way. This facilities model of the distribution system is essential for nearly every business application of system development. Knowing exactly the current state of the distribution system is critical for the operations as well. The most critical operations, of course, is dealing with power failures during emergencies. Chapter 6 is devoted entirely to emergencies. The following sections deal with nonemergency, but important, business situations of operations (i.e., running the system on a day-to-day basis).

Nonemergency Business Operations and Maintenance

The first part of this chapter dealt with the process of growing the electric distribution network, which is the capital work (or expenses related to capital work) of the distribution system. The following sections deal with how GIS supports the nonemergency daily O&M of the electric distribution network.

Maintenance

Of course, every utility must keep its equipment in good working order, so they have extensive maintenance programs. Since the distribution network has so much equipment distributed over such a wide area, it is a challenge. Utilities fix things in one location; other equipment fails somewhere else. Generally, when a piece of distribution equipment fails, customers experience an outage. So, the time and effort spent on maintenance often pays off in better reliability. However, since there is so much equipment and finite resources, utilities always have to choose what facilities to maintain and when and how often. In the past, utilities performed most maintenance on a scheduled basis, just like most people maintain their automobile. They change the oil every 3000 miles or three months, regardless of the duty on their vehicle.

However, utilities today generally maintain equipment on the basis of reliability impact. So utilities implement reliability centered maintenance (RCM). For example, the utility looks at more than just the last time they maintained a piece of equipment. They look at how often they operated the equipment, the conditions, the past performance of similar equipment, and a host of other variables. This helps them fine-tune the time when they need to service the equipment. Ideally, they will put off the maintenance longer, based on the results of the RCM algorithms. GIS provides an additional framework for RCM work, adding spatial analytics to the solution. So, for example, RCM may determine that the utility needs to maintain a piece of equipment. However, a spatial analysis shows salt contamination of that equipment due to its proximity to the ocean and a three-month history of winds blowing in from the water. So GIS spatial analysis supplements RCM. The GIS can also display the results of the maintenance process for visualization and dissemination.

Maintenance activities include the following:

- *Asset management*—this is the study of whether to replace the asset (creating a capital project) or repair the asset. Age, failure history, cost of repair, location, and condition are key elements of the analysis. The GIS provides the data management framework from the facilities model (Chapter 4) to be able to organize the work and to supplement asset management with some spatial analysis.
- *Inspection*—this is the capture of condition information. The GIS provides the guidance to organize where the inspections should occur. Often the GIS is used to collect the condition data and provide a means of updating where the condition data actually goes. Some utilities prefer to keep condition data directly in the GIS; others move the data into their EAM systems. In either case, condition data is critical to asset management, so the asset management analysis must have ready access to this data. Ideally, EAM and GIS are well integrated, so the users never need to know where the data is actually stored.
- *Testing*—testing and inspection often go together. The results of the test determine the condition of the equipment.
- *Repair*—is really a form of maintenance. If a handle is bent or a door or access panel is missing a part, the maintenance person repairs the problem.
- *Replacement*—when the field workers determine that a part is beyond repair, they will replace the part. If the equipment is a major, the field worker creates a work order and generates a capital project. If the replacement is minor, like a blown out lamp on an indicator panel or a rusted out cover, the field worker replaces the equipment directly.

Substation Maintenance

Substation maintenance involves the same activities as network maintenance except that much of the equipment is not directly modeled in the GIS or not tied to specific equipment that is modeled in the GIS. For example, utilities need to maintain substation batteries. Substation batteries are not modeled in the GIS, since they are not linked directly to any one current-carrying device but are critical for the operation of a substation.

There are two ways to deal with this. The substation building structure itself of course can be modeled in the GIS as a polygon feature. Then the GIS modeler can create a relationship to nongraphic features called substation batteries. The condition data and maintenance information would be stored in the EAM, but a spatial query could expose the location of the substation battery equipment within the specific substation. The other way is to create a specific feature within the boundary of the substation polygon of a station battery probably as a simple point feature. Then the specific batteries can be identified directly. The same process could be used for annunciators and other substation equipment not specific to a particular transformer or breaker. Other substation features, not specific to the electric operations, like doors, buildings, pumps, heaters, and lights, are best modeled as a building information model and dealt with like other buildings the utility operates.

Special Maintenance Activities

GIS provides the means to organize the work of maintenance, because the GIS helps utilities identify the priorities of maintenance activities. It shows where the field crews need to do the work. It shows where they have done the work and collects the condition of the equipment. Two additional maintenance activities require more discussion and have different impacts: vegetation management or tree trimming and street light management.

Tree Trimming

Like transmission companies, distribution companies that operate overhead lines pay constant attention to the condition of trees that are in close proximity to their lines. The best way to mitigate some of the risks associated with trees falling on wires is to regularly trim tree branches around the distribution lines. There are two types of tree trimming work: programmatic and corrective. Utilities schedule tree trimming activities on individual feeders, often based on a programmed cycle. For example, utilities will schedule specific feeders to have tree trimming once every three, four, or five years. During outages or for construction activity, utilities trim trees on an unscheduled basis often called corrective tree trimming. In many, if not most, cases, utilities contract tree trimming to specialists.

While distribution organizations often manage tree trimming, regularly shared services groups manage the overall vegetation management activities. Additional discussions of vegetation management is covered in Chapter 8.

The output product for tree trimming are GIS maps showing trimming work polygons, with imagery of the area to be managed, coded with the results of the spatial analysis showing greater or less need, work sequence, budgeted versus actual cost of the work, progress of the work, and specific tree hazards in the area. Work crews can directly enter their data into the GIS using a mobile device along with photos showing the quality of the work performed. The utility botanist can manage a greater number of contractors and work areas by accumulating the field data in the GIS, can see the progress of the work, and track the overall costs and be alerted to any unusual situations that may require additional funding.

The planning and execution of tree trimming can be significantly tuned to lower the overall costs and improve the effectiveness with the use of GIS. Figure 5.9 shows tree trimming work in GIS.

Street Light Management

Not all utilities own and operate street light systems. In some cities, like Boston, the power company owns some, the city owns some, and the park district owns others. Even customers own some of the lights. Many regulators of utilities require that utilities repair street lights within a fixed period of time from when customers notified them of a street light complaint. Most often the repair is a simple replacement of the lamp.

The process is that someone reports that a street light is dark. The utility must verify that the street light is in fact one of theirs, exactly where it is, and what type of fixture and lamp it is. Then they must record exactly when the report was made

Figure 5.9 GIS map showing tree trimming work. (Source: Clearion.)

(for regulatory reporting), and then they must dispatch a crew to the location to fix the street light.

As part of the facilities model, utilities capture the street light location and its characteristics (whether directly in the GIS or accessed from another system) within the GIS. When a utility receives a report of a street light out, they can determine the location exactly, determine if in fact they are responsible for the repair, make sure the repair person has the right replacement parts, organize all the work nearby to optimize the work, then make the repair, and finally report exactly when they made the repair.

Many utilities offer a GIS-based online or mobile app for reporting of a street light outage. Using GIS, the person making the report can easily identify which light is out and the utility can automatically determine the ownership and responsibility for the repair. In fact, the app can route the repair order immediately to the lamp repair person directly. In a non-GIS-centric street light repair scenario, the utility will often ask (online or over the phone) what the pole number is of the street light in question. Since in the vast majority of cases, the report of street light outage occurs at night, and with the street light out, it is probably too dark for the customer to read the pole number. If instead, the customer could simply click on the street light location in the GIS, enabled in the smart phone, they specify the exact location, without having to know the pole number. They can also add other data, like damage to the pole or to the street fixture.

A self-service-driven GIS-enabled street light application shortens the entire process of reporting, checking records, dispatching, and repairing the lamp. Other than the actual replacement of the lamp, the GIS entirely automates the process. See Figure 5.10 for an example of GIS-enabled street light application.

Utilities can also take the street light outage reports and use them in conjunction with crime statistics to help law enforcement correlate frequent street light outages with crime activity. Rapid repair of street lights in high crime areas can provide a significant community benefit.

Nonemergency Operations

In addition to power failures, utilities have to main adequate levels of power quality. This involves avoiding flicker, high and low voltage, excesses losses, and momentary interruptions. Utilities can help mitigate these issues by identifying the sources and measuring the results.

For example, flicker, or the rapid changing of voltage, which causes noticeable changes in the lighting (that is, "the lights are flickering") is often caused by specific kinds of loads, particularly arc-welders. The GIS can identify where these offending loads are and then craft solutions to mitigate their impacts.

Voltage variations are usually a result of operational situations that have changed since the initial design of some portion of the network. For example, if customers connected to a particular transformer and secondary main over the years have increased their loads significantly, then they are apt to experience low voltage. Customers adding additional air conditioning, high-definition TVs, or extra refrigerators, could drive the voltage lower than what the system can handle. Should the customer add an electric vehicle charger, that might well create a low-voltage situation, particularly if the customer is at the end of a secondary main.

Momentary Outages Are Annoying

Momentary outages occur during switching events and during power failure restoration events. The common medium-voltage recloser is a circuit breaker that when it senses a fault condition, (i.e., when it senses that the current exceeds normal

Figure 5.10 GIS-enabled street light application. (Source: Esri et al.)

values, most often during a short-circuit situation), it trips the breaker, causing a power failure. However, a timer in the recloser closes the breaker allowing current to flow again after a short interval. If the recloser senses high current again, it opens again and stays open. It could also be set to perform this open and closing several times. The idea is that if a tree limb falls across a line or a squirrel's tail shorts across a transformer's bushing, the short circuit current occurs only for a short period of time. After the fault, the tree limb or dead squirrel falls to the ground. When the reclosers closes back in, if the source of the fault is gone, the recloser restores the power and customers do not experience a lengthy power failure.

All this is positive, except that this opening and closing of switches creates a momentary outage, which could create problems for customers. Digital clocks stop working. Maybe computers lose files or the situation causes data corruption. As networks get smarter, these momentary outages in fact may become more frequent. As noted in Appendix B, smart grid provides this concept of self-healing, which means that the electric network will attempt to configure itself to minimize total customer outage time. In effect, total outage duration will drop, but the number of momentary outages might actually increase.

Nonetheless, utilities have to manage power quality issues. They do this by monitoring and measuring what goes on in the network. Utilities measure the total impact on their system of momentary outages using a metric called momentary average interruption frequency index (MAIFI). The definition of a momentary outage is somewhat vague and varies around the world. It is a measure of outages that lasts for less than a minute. Some utilities use 5 minutes. The calculation is the total number of times a momentary outage occurs during a calendar year divided by the total number of customers served in that area. In effect, it is the measure of how many times per year the average customer experiences this kind of disturbance, regardless of the cause. The cause could be lightning, switching, or reclosing. If the momentary outage is the result of a planned event, like the cutting over of one line to another, the utility will not include that event in the MAIFI calculation, even though the customer may not know the difference.

Often utilities report a single MAIFI number for their entire service territory. A typical MAIFI in the southeastern U.S. is around 10. The problem with that number is that it averages over such a wide area. Using GIS, utilities can plot momentary outages over a much smaller resolution of service territory. So while the average of MAIFI might be 10, there may be areas of MAIFI of 1 or 0 and another of 100. Utilities can use GIS to continuously monitor MAIFI and be able to uncover where their areas of poor power quality lie. Then they could analyze the areas more closely to determine if tree trimming needs to be increased (or decreased) or whether they need to examine the types of loads, like sensitive manufacturing, to help customers deal with annoying power quality events. GIS not only provides a visualization of the problem, in other words, shows where the problems are most severe, but allows utilities to correlate the problem in a more effective way. Figure 5.11 shows MAIFI on a GIS map by zip code.

Transformer Load Management

Utilities rarely measure the loading on distribution transformers. So today, the biggest mystery of the electric distribution network is the past, current, and future

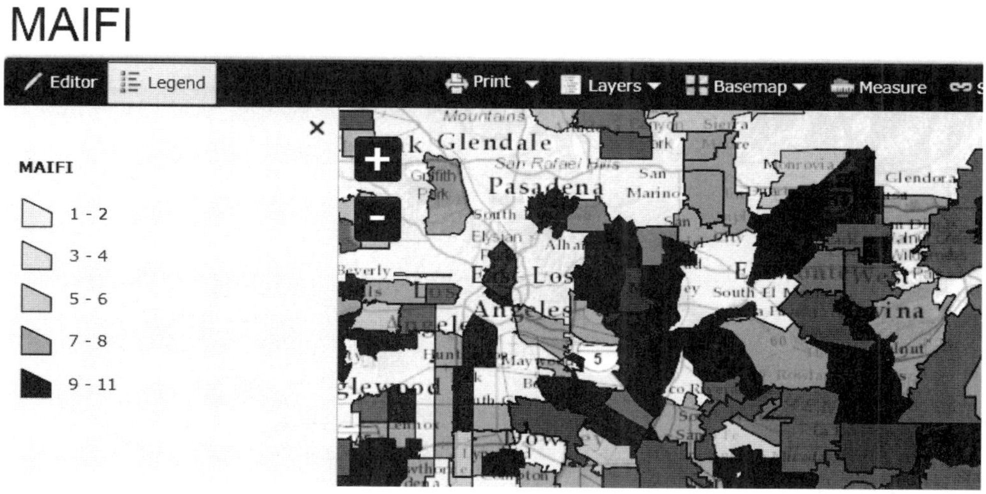

Figure 5.11 GIS map showing MAIFI by zip code. (Source: Esri et al.)

projected loads on distribution transformers. The result of this lack of knowledge is unexpected failures of distribution transformers particularly during periods of heavy demand. This can be particularly problematic during hot and humid evenings when crews are off shift. Customers are out of power for long periods of time without air conditioning.

Without smart meters, utilities have to estimate the load on each distribution transformer during a projected heavy load period. Short of monitoring the loads on the transformers in real time, the utilities do not have an effective means of calculating coincident peak loads. They have to take billing data from the customer information system and apply some sort of estimated load pattern to the billing data to estimate the peak load. They then have to determine the coincident demand of all the customers connected to the same transformer.

Smart meters and GIS solve this problem. With smart meters, utilities can determine the exact customer load at every time interval in the day. If the utility created a good low-voltage model in the GIS or at least linked customers to the transformers in the GIS, then the utility can calculate the total coincident loading for each of its transformers for each interval that the utility measures consumption. They will then know which transformers have experienced excessive loading and which have experienced loss of life. The GIS can display which transformers are vulnerable to failure.

Utilities rarely monitor each smart meter in real time. Smart meters will typically store interval consumption data, typically from 5 to 15 minutes and then transmit the data several times during the day. So while the utilities will not know exactly when a transformer is overloaded in real time, they will be able to know exactly what the transformers experienced several hours later or at worst, the next day. In the past, they would have no knowledge of transformer coincident loading.

When utilities project the load to increase, they will know which transformers are most likely to fail and take action to mitigate that failure. They can use GIS to determine which transformers to upgrade or which areas of the network to apply load management or demand response to. Rather than shedding load unilaterally, utilities could shed load to only those areas that are at risk of losing a distribution transformer. This might mean a tradeoff of a 15-minute outage for an extended outage to replace a failed transformer during a particularly busy time.

This illustrates how GIS helps utilities narrow the problem down to smaller areas to help prioritize issues to avoid problems in the future.

Call Before You Dig

Most jurisdictions require utilities to participate in a program variously called "call before you dig," "dig safe," and "safe dig." An agency within a state, a region, a province, or a country coordinates all digging within the jurisdiction. In the U.S., each state has such an agency. The rule is that if anyone—a contractor, a home owner, a utility, or the municipality—needs to dig a hole, they must first notify the call before you dig agency of their intention to dig and precisely locate where the dig is to occur.

The workflow is a utility receives a request from the call before you dig agency and determines if they have underground facilities in the area. Then, if they do, they must mark the area where their facilities are located and notify the agency that there is an underground facility that they have successfuly marked in the field.

Some utilities set up a self-service system in which the call before you dig agency can directly query the GIS. Then, if the query results in a discovery of an underground facility, the GIS creates the marking work order, which includes a GIS display of the facilities as part of the work order. The GIS-based work order automatically delivers the information to the field personal on a laptop or mobile device with details on where to mark. Once complete, the field technician completes the work order in the field on the mobile GIS device, closes the work order, and notifies the agency. The information products are the original request, the map showing the request, and the discovered underground facilities. The self-service process closes the work order immediately if there are no underground facilities in the area.

Third-Party Attachment/Use Management

Entities such as wireline, wireless, cable, and fiber-optics companies regularly use electric company structures such as poles, towers, and even buildings for their equipment. Municipal fire departments often attach fire alarm equipment and cables to electric poles. It is not uncommon that over time, utilities that own the poles lose track of exactly what equipment other companies have attached to their poles.

Utilities almost exclusively use GIS to document these third-party attachments and manage a workflow to calculate the various attachment fees. Utilities can use GIS to facilitate the collection of these attachments should the utility wish to perform a full audit. Some utilities find that the lost revenue from collection of these third-party attachment fees is substantial. The information product consists of a simple map illustrating the electrical feature, like the poles with symbology of the various attachment types shown. GIS can perform a simple count analysis for each region identified on the map. For example, the GIS can determine the number of fire alarm boxes within a town boundary and the length of the fire alarm circuit. They can display this as a table or chart on the map, and they could update the display whenever a third party adds, edits, or removes an attachment.

It is also not uncommon for utilities to lease space within their electrical duct bank system. As detailed in Chapter 4, many utilities have built underground networks of duct banks, manholes, and conduits traveling up and down the city streets and into and out of all the basements of all the buildings in the city. If the utility has unused pipes, it can lease the space to other companies, like fiber or cable companies. The GIS is the most common tool to manage these cable locations and be part of the workflow to render bills to companies leasing the ducts.

Another important consideration is that the utility may have to notify the third party should damage occur. If, for example, an electrical cable faults in a manhole causing a fire to erupt, the utility will need to know not only which electrical equipment may be damaged, but also what third-party equipment may be damaged. The GIS provides this if there is a good process to keep this data updated.

GIS Used in Electric Distribution Development and Operations

Electric distribution departments were the first to embrace the notion of digital mapping, mainly to solve the messy process of constantly updating their distribution operations maps. Many built digital mapping systems that solved only the map production process. Over time, utilities realized that the information they maintained in their GIS had enormous value for more than just creating clearer and easier-to-produce maps.

They realized that they could use GIS technology to better plan for the development of their electric distribution network by fine-tuning the process of load forecasting. They could use GIS to actually design the network itself and communicate the designs to the construction crews and contractors. They could use the GIS to manage street lights, call before you dig requests, third-party attachment and power quality issues, transformer loadings, and smart grid deployments.

In many parts of the world, falling trees and tree branches cause many of the outages on the distribution system. Often the cost of tree trimming is the third largest expense behind labor and material for a distribution utility department. The GIS can organize and analyze the work to optimize tree trimming activities by bringing in relevant data (like historic rainfall) from outside the company and combining the data with information within the utility to produce a more intelligent and cost-effective plan.

Finally, the GIS is essential to mitigate the risk of failure, help ready the system for weather events, help in the restoration of power, and provide assistance in the rebuilding of the network after a major outage event. Chapter 6 will cover this in detail.

CHAPTER 6

Emergency and Outage Management

Emergency Management

An electric utility's primary mission is to keep the lights on at all times. So when the lights do go out, a utility's single most important activity is to get the lights back on as quickly as possible. Society rates how good an electric utility is by how reliable it is. So, whenever there is an unexpected or unscheduled power outage, utilities execute a workflow to find out what caused the problem, assess the damage, and fix the problem. Utilities have been doing this since the day Edison turned the switch on Pearl Street Station in New York City more than a century ago. Of course, location and mapping play a critical role in the workflow of doing just that.

There are two types of power failures that result in two very different responses: the major emergency and the routine outage. The difference is stark. There is no real hard and fast rule that differentiates a major emergency outage from the routine, but generally if the majority of customers served by a utility are out of power, that is a major emergency outage. While a large power failure is an emergency, if the existing utility staff can restore the outage using standard processes, then utilities consider this routine. The media may report the outage, but once they do, most people not out of power quickly forget about it. Most power failures are routine and for many utilities occur every day. The restoration process is fairly predictable. However, the quality of performance of utilities to manage this process varies widely. Accurate information, timely communication, and a high degree of collaboration of all parties involved often determines how well a utility deals with power failures.

GIS plays a major role in a managing the information about the location, characteristics, and condition of the utility assets, and provides a framework for communicating the state of the electric network. It also provides an effective way of getting all workers on the same page—in other words, of facilitating communication. After all, every utility uses maps to pin-point places where problems exist. However, unlike in the old days, when people pinned hard-copy maps to a wall, GIS delivers the map and the events electronically to everyone in the company on whatever digital device the workers happen to have. Of course, that's the ideal. Today many utilities still use old processes to deal with outages. Figure 6.1 shows the differences between major outages and routine events.

Routine Outage

- Existing staff handles
- Limited damage
- Central dispatch
- Union work rules
- OMS used extensively
- Outages included in SAIDI, SAIFI

Major Outage

- Foreign crews
- Extensive damage
- Decentralized dispatch
- Work rules suspended
- Limited OMS use
- Outages excluded from SAIDI and SAIFI

Figure 6.1 The differences between a major outage and a routine outage.

How Well Does the Utility Manage Its Reliability?

Utilities have come up with standard ways of measuring reliability. Nearly every utility measures the total accumulated time that an average customer is out of power over a year. The industry uses the term *system average interruption duration index* (SAIDI) as a way of comparing reliability peformance of electric utilities.

SAIDI equals the total number of minutes of power outage experienced for every customer for the year divided by the total number of customers.

If a 100,000-customer utility experienced 12,000,000 minutes of total customer outage, SAIDI would be 12,000,000 divided by 100,000, or 120 minutes. Of course, of those 100,000 customers, perhaps only a fraction actually experienced any power failures at all, while some may experience many hours of power outage many times. So SAIDI may not resonate with customers, since it is just an average. In the U.S., 120 minutes or 2 hours is about the average of all utilities. This value varies widely around the world. In parts of the world, SAIDI is below a minute. In other areas, SAIDI can be multiple hours, days, or even longer. Despite its flaws, SAIDI is a good indicator of how well a utility operates and its commitment to a culture of restoration. It would not be uncommon to find SAIDI as a key performance metric at the highest levels of management in a utility company.

SAIDI doesn't tell the whole story of response to emergency or routine outages. Often the calculation of SAIDI excludes some events, like major storms or other natural disasters. Some utilities exclude outages lasting less than 5 minutes in the calculation, since they include these short outages in the calculation of MAIFI (see Chapter 5). Some utilities report SAIDI with and without abnormally large events.

There is no standard measure of outages for those events excluded in the calculation of SAIDI, although the Institute of Electric and Electronic Engineers (IEEE) has developed a metric to measure these abnormal or nonroutine events [1].

Other Reliability Measures

SAIDI measures the total accumulated time of an average customer to be out of power during the year. System average interruption frequency index (SAIFI) measures the number of times an average customer is out of power. Like SAIDI, SAIFI probably wouldn't resonate with any customer.

SAIFI equals the total number of power failure occurrences over a year divided by the total number of customers.

Like SAIDI, the calculation of SAIFI often excludes major storm events and outages that last less than five minutes. An additional metric is customer average interruption duration index (CAIDI), which calculates how long on average a customer experiences an outage.

CAIDI equals the SAIDI divided by the SAIFI.

Since CAIDI is a derived metric, it is less frequently used. Thus, the major metrics for reliability are SAIDI, SAIFI, and MAIFI.

Reliability and GIS

Aside from the varying ways utilities measure SAIDI, SAIFI, and MAIFI and what they include or do not include, they are averages and are not particularly good proxies for what customers actually experience. Utilities have attempted to get a better handle on the values by assigning these metrics to individual medium-voltage feeders. In this way, they can determine which feeders are more reliable. However, higher-voltage medium-voltage feeders can be quite long and cover vast areas, so this practice might still not be a good indication of how customers feel about outages. In addition, unless utilities are able to associate customers to feeders at all times, they won't be able to determine an accurate reliability ranking for feeders. During outages, for example, utilities will regularly configure sections of feeders to provide alternate paths. More utilities will implement automated self-healing processes, which will involve automated switching of remotely controlled switches during outages or periods of heavy loading. This means that any number of feeders in all kinds of configurations may supply customers. Thus, customers assigned to specific feeders will become less common. The notion of self-healing is a characteristic of smart grid.

GIS, coupled with smart meters, provide a fuller picture of the customer experience of reliability than simply relying on system averages. With smart meters, utilities know exactly when customers lose power and when the utility restores power. During a single event, for example, a customer may experience several periods of power outage. Utilities would know precisely the duration of any single event; thus, their calculation of MAIFI and SAIFI would be completely consistent. Without smart meters, utilities cannot accurately represent every instance of a power failure.

Since GIS knows the location of every customer, it can simply map areas of reliability. It can map by block, zip plus four, or any other spatial grouping. The utility can then use the commonly used terms like SAIDI, SAIFI, and MAIFI but base the results on much smaller areas. Then the utility can use the areas of high SAIDI (high is bad) and intersect those areas with the distribution infrastructure within that area. In this case, the utility would expect to see wide variations from one area to the next and be able to fine-tune their reliability program to areas rather than to long medium voltage distribution feeders or substation regions. Further, the utility could better understand the underlying causes of why exactly one area is more reliable than another. With that information in hand, the utility could craft a reliability program that focuses on the very worse areas.

Using spatial analysis, the utility can correlate voltage levels, equipment age, types of vegetation, weather, and tree-trimming history to discover the underlying cause of the reliability issues. They may want to obtain additional information, such as motor vehicle accidents or winding roads in the area, or that this neighborhood is at the end of a long feeder or is subjected to salt contamination or a variety of factors that could answer the question of why the reliability is so bad in this area. The GIS would be able to illustrate the average interruption frequency and duration over a relatively small area compared to averages over the entire service territory. Figure 6.2 shows a GIS map with SAIDI rankings by zip code.

In addition to providing accurate network data, the GIS can help focus the utility on where it needs to improve.

Risk Profiling

There are many elements of risk for the distribution business: the risk of overspending, of employees getting hurt, or of damaging the environment by having oil-filled equipment rupture. Perhaps the most common kind of risk for an electric distribution business is the risk of equipment failure creating a power failure. While utilities cannot control the weather or earthquakes or wildfires, utilities can determine their vulnerability to these risks. This section will create one of the illustrations of the power of GIS spatial analysis to determine the risk profile of a distribution network to failure and of power loss.

There are two elements of risk:

- The likelihood of the failure itself;
- The consequences of the failure.

Together they determine the relative risk. For example, the likelihood of an overhead line being taking down by a falling tree branch depends on whether the

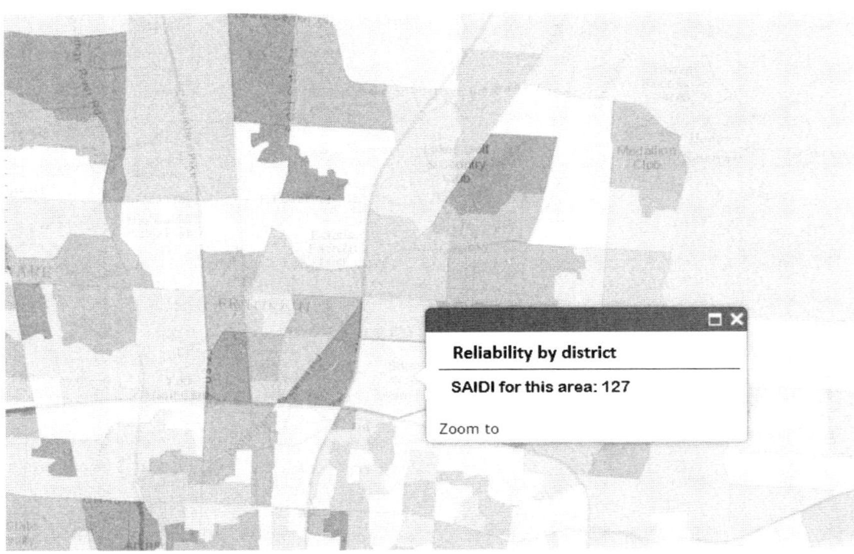

Figure 6.2 GIS web map showing areas of SAIDI by zip code. (Source: Esri at al.)

line is near a tree, the type of tree, when the nearby tree was last trimmed, and how likely the area is subject to severe weather. It also depends on how strong the wire is, the type of construction, and a host of other factors. The consequences of the damage depends on whether the section of line is at the end of a long feeder or at the beginning, or if the feeder supplies critical load and if the area is easy to get to or on the wrong side of washed out bridge on an island, for example. Other consequences can be revenue lost, or a large outage possibly caused by damage to the feeder. Perhaps the damaged feeder has had a number of prior failures; the utility may consider that section a high-risk section. If the feeder supplies prominent members of the community, the loss of that feeder section could have political consequences. What if the feeder supplies power to a hockey rink during the playoffs? The GIS helps to illustrate the risks in an orderly and scientific way.

GIS Tools for Risk Assessment

The more information that is available in the risk assessment, the better able the utility is to determine risk, and consequently the better able the utility is to mitigate or at least prepare for the risk. If a distribution utility knew that a certain line was particularly vulnerable to 75-mile-per-hour winds, then they could harden the line section or at least have crews ready to repair the line should those winds materialize. Regulators judge utilities not simply on the speed of restoration but on the degree of planning and analysis of outage events.

Equipment failures often come as surprises to utilities. For example, during hot summer months, transformers can become overloaded and fail, often to the surprise of utilities. Yet, utilities have information that could determine if a transformer or other device is vulnerable to failure if they perform a comprehensive risk assessment. They can base much of the information for the risk assessment on location in combination with other data readily available in the utility and over the web.

The following example illustrates the use of spatial analysis for a storm planning risk model. The idea is for the utility to create a dynamic model that queries a variety of factors, many of them from different sources, normalizes and weights the factors and presents the results as a GIS map showing the relative risk of an overhead electric distribution system to a storm event. They can perform the same type of analysis for vulnerability to a high heat event or a flooding event.

Vulnerability Factors

Recall that risk involves vulnerability and consequence. So the first step is to identify the vulnerabilities of an overhead electric system to a possible storm event. In this case the storm event is high winds; say, more than 75 miles per hour. Data sources include the following:

- *Map of historic high wind events over the last many years.* The idea here is that the baseline for vulnerability of the event is the relative likelihood of the event from even happening. So a data source is a map that shows polygons of the occurrence of high wind events. The map would show polygons of frequencies of storm events having greater than 75-mile-per-hour winds over some historic period. The map is a raster image with each polygon having a uniform value of likelihood of storm event—say, in a scale of hours per year on average

- *Map showing age of conductor.* This map would show each conductor segment highlighted by some age interval, such as 10 years or newer, 10 years to 25 years, and older than 25 years. The data source for this map would come from the company's work management or plant accounting system.
- *Map showing pole inspection data.* Here the condition of the poles would be plotted based on the last time the poles were inspected. For simplicity, the map data would have several levels of condition, such as poor, fair, good, or excellent or have some scale of 1 to 5 or 1 to 10, depending on the condition.
- *Map showing age of poles.* This data would come from the company's material system or asset system. Similarly, the map showing the pole age would be bracketed into a few intervals.
- *Map showing soil types.* This map details the soil configuration from clay to sandy over a few bracketed intervals. The idea here is that poles are more apt to topple if they are buried in sandy soil. Also trees will also be less stable.
- *Map showing feeder tree-trimming history.* Falling tree branches cause the majority of outages during wind events, so knowing when the last time trees were trimmed gives a good indication about the chances of branches falling on lines.

Each of these maps of course are digital maps from internal or external sources. The utility's GIS would likely access these maps as web services from the organization that manages this kind of data. The above data sources are just samples of data that goes into the vulnerability analysis. The utility could add other factors such as past rainfall, type of vegetation (pine trees versus hardwoods), elevation, line maintenance, guying, and pole loading, among others, into the analysis.

The GIS vulnerability model accesses all these data sources from the core systems that maintain them. For example, soil type data might arrive into the analysis as a web service from the local conservation commission. The idea is that every time the utility runs the model in the GIS, the model accesses the data sources from the most recent available source. If the utility recently replaced a pole and documented that replacement in the work management system, the next time the utility ran the model it would use the new pole information.

Since each of the data sources may be coming from different systems with different map projections, the first step is to convert all the data sources to the same map projection. The GIS performs this by examining the metadata, or data about the data, to determine the map projection of the source data and if necessary converts the data to the target projection, most likely of the company's electric distribution GIS.

The next step is to convert the vector data, such as the map that shows conductor age as a line segment into a raster data set. In this illustrative example, the model converts all data sets to raster images.

The next step is to normalize each raster data set to a common scale of values. For example, age may be bracketed into a scale of one to ten, with ten being the newest, or age may be displayed explicitly as the actual age of the conductor. All the data sets need to be converted to the same scale. While is it interesting to view many data sets displayed together as layers, it becomes difficult to extract real meaning from these layers if there are too many of them.

The next step is called weighted overlay. This spatial analysis tool was illustrated for transmission routing in Chapter 3. Since all of the layers now are based on the same map projections and have the same relative scales of vulnerability severity, the GIS can add up the various values in each raster cell to create a new map that displays the total vulnerability of each raster cell. However, the model can include logic whereby each data source can be given a different weighting factor. So, for example, in the vulnerability assessment, utilities may believe that historic rainfall or soil type has lesser or greater importance than pole condition, so they can assign a weighting factor to one or more of the data sets.

The resultant map is the total vulnerability to a wind event based on many factors. In fact, a utility could even include a map layer created by experienced distribution workers that outlines areas that they think are most vulnerable without regard to any scientific data, just their own intuition or experience. As every insurance industry expert will tell you, the more information you have, the better handle you have on vulnerability. The final map shows a heat map (i.e., those areas whose combined scores of vulnerability fall within a certain range are shown in a different color or shade). Normally the darker the shade is, the higher the vulnerability.

Consequence Analysis

Once the GIS creates the vulnerability map, it can create a similar consequence map. This can be done in a number of ways. Utilities can assign each feeder segment a value based on the criticality of the feeder section, such as the number of critical loads served by the feeder section or the total number of customers served or perhaps areas that are hard to get to. Planners can then perform a similar weighted overlay analysis to the vulnerability analysis by taking all the consequence factors creating a single consequence map.

Total Risk

The total risk profile is the combining or mashing up of or simultaneously displaying of the vulnerability map with the consequence map. Once the GIS creates the total risk map, it can be run regularly as situations and the feeder data sources change. To test for electrical loading, the analysis can include real-time data from SCADA or DMS to be able to perform more ad hoc risk profiles during outage events.

Of course, once the utility maps the total risk, they now have the tools to be able to perform risk mitigation, which may include replacing key poles in areas of high risk or staging replacement poles and other equipment near areas that are more likely to fail. A proximity analysis could be used to answer the query: show me the closest parcel of land owned by the utility or available for lease to the areas of high vulnerability large enough to store emergency equipment. The spatial analysis could compare the various lease costs to do this. See Figure 6.3 for the steps involved in a GIS risk analysis.

Figure 6.4 is a GIS map of the results of a risk analysis for a wind storm.

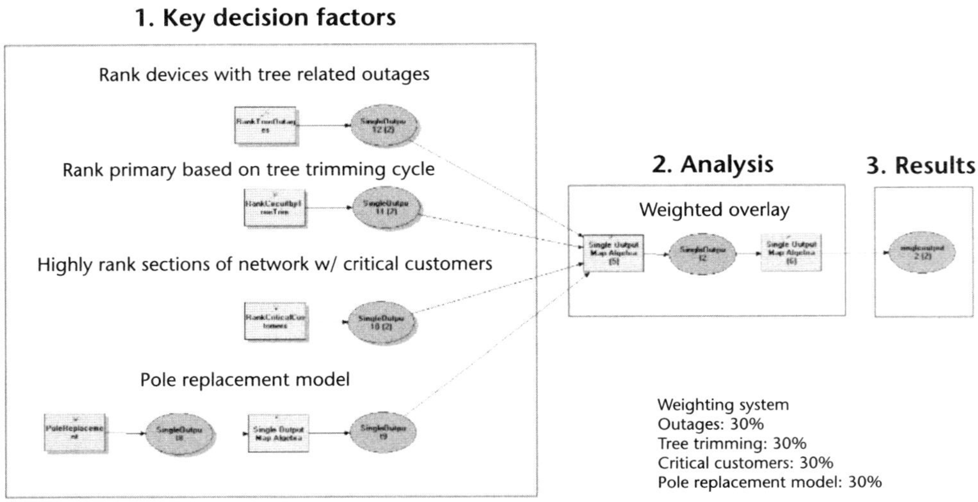

Figure 6.3 Steps in the spatial analysis for a GIS risk analysis. (Source: Esri et al.)

Figure 6.4 GIS map showing the results of a risk analysis for a wind storm event. (Source: Esri at al.)

Emergency Management for Major Outage Events

When some very serious emergency or disaster strikes that results in significant numbers of customers without power, utilities often initiate some form of major emergency plan. These plans will usually differ considerably from the workflows that utilities use for routine outages. As stated earlier, there are no hard and fast rules that determine when a utility in effect crosses the line from a routine outage to a major outage. Often utilities make the assessment that if they cannot restore the vast majority of customers within one day with their existing workforce, they declare a major emergency. This declaration often determines if the customer outage is or is not included within its SAIDI numbers. Since utilities must report SAIDI (and SAIFI) to regulators, they want to be sure that the emergency is justifiable, since they do not want to be accused of calling an emergency without cause, just to make the numbers look good.

The most common form of emergency is, of course, bad weather: blizzards, tornadoes, hurricanes, floods, and ice storms. Other less common emergencies are wildfires, earthquakes, and mudslides. Emergencies can be nonnatural, such as cyber attacks, terrorist, wars, and vandalism. Even cycles of drought and heavy rain can cause underground pipes to break, causing widespread damage and outages. People even theorize that sun storms could cause electromagnetic equipment like transformers to fail, which would cause widespread blackouts.

Major Emergency Plans of Operations

As noted when these major events occur, utilities often execute a completely different workflow for restoration. Some of the elements of the plan include decentralization, establishment of temporary local restoration centers, importing crews from other utilities, suspending union work rules, and using office workers for damage assessment.

When utilities decide to decentralize operations, they may relocate distribution dispatchers to the local service centers. Of course, this leads to a greater need for communication and collaboration. During the critical periods when the event is still under way, outage management systems are not particularly effective, since their prediction engines depend on customer calls (or smart meter feeds) to help them determine the location of damage. When there is widespread damage, when everyone is out of power or many medium voltage feeders have tripped out, an OMS does not provide knowledge about the extent of the damage. Decentralization is an attempt to break the problem into manageable sizes. The utility calls upon other utility employees who have normal "day jobs" to help with the logistics of restoration during major events and to establish regular local media and political contact processes.

In cases where an emergency event causes significant damage, the utility must establish an enhanced process for materials management.

Utilities will often establish local restoration centers, in shopping malls, schools, and police and fire stations. These facilities often serve as local control centers and places where restoration officials house and manage temporary employees and foreign crews (crews from other utilities on loan).

Once a utility realizes that the impact of the event is beyond their ability to adequately restore with in-house facilities, they will regularly contact other utilities. They normally have mutual assistance agreements in place. This action has serious financial and logistical consequences. Further, timing is critical. If, for example, the event is a fast-moving hurricane, causing widespread damage, getting assistance from any utility that either has been hit with the hurricane or expects to be hit is extremely unlikely. So the utility will have to ask for assistance outside the area the storm is hammering. However, other utilities in the region are also looking for crews or a commitment to send crews.

It becomes critical for the utility to properly assess the need for crews early enough to get them deployed and not so early as to have the crews show up and not be needed.

The cost of these foreign crews, of course, is borne by the requesting utility. Once the impacted utility requests mutual assistance, they must manage those resources. This includes providing food and housing. The utility must be careful to keep good records of all the costs incurred by foreign crews. They also must be ready to provide meaningful work for the crews when they arrive.

Damage Assessment

Perhaps the most critical workflow for the utility during and right after a storm event is the assessment of exactly what the damage is. During routine outages, utilities call upon shift troubleshooters to assess the damage, and in many cases they actually fix the problem causing the outage. When the damage is extensive, they need to call upon other employees, the day job employees or foreign crews, to assess the damage. Many utilities still use relatively primitive tools to capture that knowledge. While it is important to know how many customers are out of power, it is more important to know the exact extent of the damage to its equipment. Having an accurate and early assessment of the damage helps utilities decide just how much help and materials they need. It further helps them to set the proper expectations to the community as to exactly how long the outages will last. GIS in the field can provide an accurate method of pin-pointing damage to the electrical facilities to exactly where the problem exists, like forms, reports, and scribbles on yellow pads can't. Figure 6.5 illustrates a GIS map showing the location of damaged facilities.

Utilities that fail to properly assess damage and set realistic expectations often find themselves in after-the-event audit sessions with regulators.

Labor Issues

Accurate prediction and damage assessment avoids labor issues during and after the event. Many union contracts stipulate that the company and union can suspend some work rules during periods of a major declared outage emergency. So the utility needs to notify the collective bargaining unit of exactly when they suspend the rules and when they declare the emergency over. Of course, this means that union leadership will be looking for assurances that the event actually warranted being called an emergency and will want assurances that the company did not keep the emergency active after the situation is back to a relatively routine state. So damage

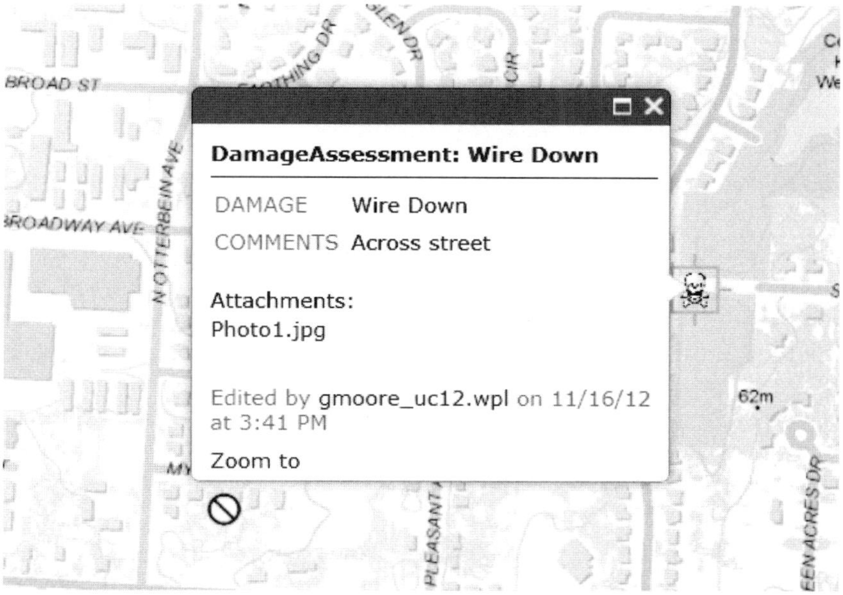

Figure 6.5 GIS in the field used for damage assessment. (Source: Esri et al.)

assessment, prediction, and communication are critical to maintaining transparency with the unions all facilitated by GIS.

Tools of the Emergency Trade

Utilities have been dealing with major emergencies since the beginning of the industry. Since emergencies are relatively infrequent, utilities often do not implement special automation and information technology equipment just for special circumstances like major emergencies. Also, emergencies are unpredictable, so it would be hard to design a system to handle every conceivable type of emergency. The result is that most utilities resort to procedures that have worked in the past. Many of these tried and true procedures are manual and paper based. So utilities will use printed maps. Many will print map books and wall maps for distribution to their decentralized emergency centers. Some still resort to clip boards, pencils, and yellow pads.

They will use their conventional automated systems, like their customer systems, outage management systems, and SCADA. However, these systems quickly get overwhelmed if the vast majority of customers are out of power. During a major power failure, utilities do not need to figure out who is out of power. They already know: everyone is out of power in a widespread area. SCADA tells them which medium-voltage circuits or transmission lines are out. They already know that. So, while helpful, the tools the utility uses during "routine" outages are much less effective during a major event.

Instead utilities rely heavily on weather and news services, TV, phones, their radio systems, spreadsheets, police and fire calls, government agencies like FEMA, the media, and calls from volunteers—and forms and more forms. They also hear from disgruntled customers. Rarely have utilities truly automated their emergency management program in a comprehensive way. What follows is how GIS can help create a holistic approach to electric utility emergency management, without having it linked to a particular type of emergency.

The Four Rs of Emergency Management for Electric Utilities

Emergency management begins well before the threat of an emergency and ends well after the lights go back on. There are four major components of electric utility emergency management: the four "Rs." They are *risk mitigation*, *readiness*, *response*, and *recovery*. Each of these components or phases of emergency management deals with location. Figure 6.6 illustrates the activities in each component of emergency management.

Risk Mitigation

As noted earlier in the chapter and throughout this book, the GIS is a critical tool to help determine risk. Chapters 2, 3, and 4 illustrated how GIS can lower the risk of building a power plant, a transmission line, or a substation in an unsuitable location by taking a number of data sources, performing spatial analysis, and displaying suitable locations. Not doing that increases the risk that a utility will build the plant, transmission line, or substation in an area that may be more vulnerable to damage. Chapter 5 detailed how utilities use GIS to manage a number of risks,

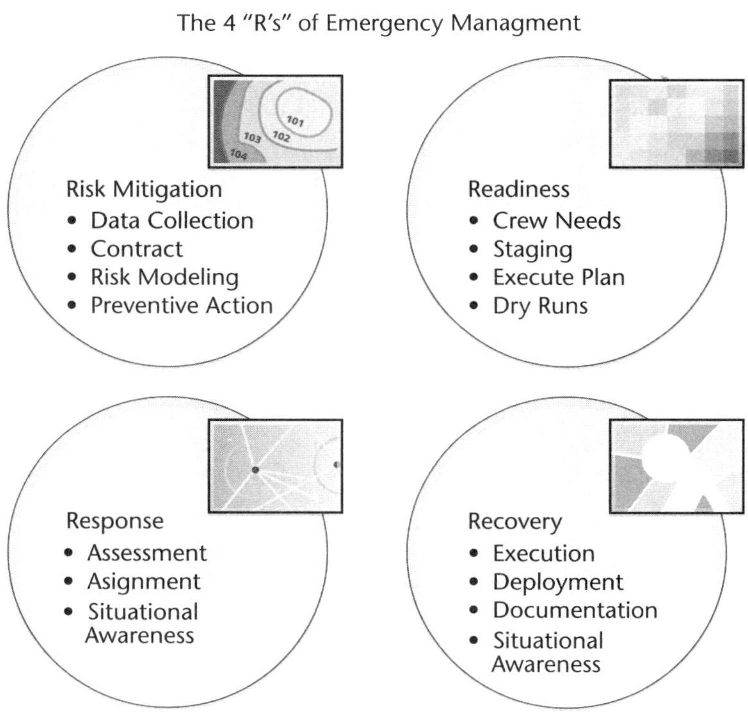

Figure 6.6 Activities during the four Rs of emergency management.

including the risk of digging into a distribution line. Risk mitigation as part of emergency management deals with looking at various threat scenarios and identifying which parts of the system are most vulnerable. With that information in hand, the utility can at least know in advance which parts of the system are most vulnerable to specific kinds of events. For example, an early winter storm in the Northeast will impact areas when leaves are still on the trees. Flooding an area of trees with shallow roots creates high risk of falling trees. Modeling this in the GIS allows utilities to do a better job of mitigating some of the risks. Certainly knowing that a major flood event may inundate a substation well in advance of bad weather gives the utility the time to install barriers, pumps, or simply to have plenty of sandbags at the substation.

GIS and Risk Mitigation and Planning for All the Disasters

The GIS provides the ideal framework for modeling the impact of a variety of emergency events on the electric infrastructure. Or, the GIS can help utilities in the site selection for new facilities by minimizing risk. So risk mitigation activities involve knowing where vulnerability is greatest and planning new facilities to minimize vulnerabilities. The GIS can model the worst-case scenarios.

Mitigation of risk can be relatively straightforward, like increasing tree trimming in those areas where trees are most likely to cause the most damage or repairing and clearing transmission access roads in critical and vulnerable areas. In areas where an important transmission line section is not easily accessible, for example, utilities will at least know what they would need to do should those transmission line segments fail.

Once a utility has a solid understanding of the risk, using standard GIS risk models and implementing mitigation measures, they will be in better position to create readiness when the real thing approaches.

Readiness

The next component of emergency management is readiness. At this point, utilities will know what the threat is. The threat is imminent. A hurricane is barreling up the coast or a blizzard is within hours of striking. Of course, some events, like earthquakes, which happen with no warning, require utilities to be in a constant state of readiness.

Utilities that have modeled an event in advance will have a better chance of predicting the event's impact and will be able to act in a timely manner to ready itself. As stated earlier, one or two of the most difficult decisions a utility emergency management officer has to make is to declare the emergency itself, since this sets in motion a host of other activities. The second decision deals with foreign crews: when to order and how many. With good data and accurate prediction models performed well in advance, the utility will have a solid understanding of how bad the event is going to be. If they wait too long, no crews will be available. If they jump too soon, the utility may end up with too many crews (if the event doesn't materialize) and a costly bill. Utilities have struggled with this for years.

GIS and Readiness

The tasks in the readiness phase includes the following:

- *Weather forecasting and weather's impact on the facilities and the surroundings.* If utilities have built vulnerability models in GIS, they can mash them up with the weather predictions. Critical parameters include predicted wind speeds, flood depths, icing, landslide intensities, and fire directions.
- *Staging.* Once the utilities have a solid prediction of the potential impact of the event, they can stage people and material where the models predict the most damage, rather than waiting until the event is over to do the staging. It is often too late to do adequate staging after the event and certainly not during the event.
- *Logistics.* Utilities can build in plans for the movement of people and material during the readiness phase and during the response to the storm. They can identify where roads and bridges are likely to be damaged and take that into effect well before they find out the bridge is out or the road is closed. If the model is done in advance, they will have much more data to rely on, versus having to do extensive field assessment of damage.

The critical readiness phase occurs during the "calm before the storm," when danger is coming, but there still is enough time to become ready, without over- or underreacting. Then when the event hits, the utility will have a solid plan for response. Figure 6.7 shows a GIS map with areas identified where crews should be staged.

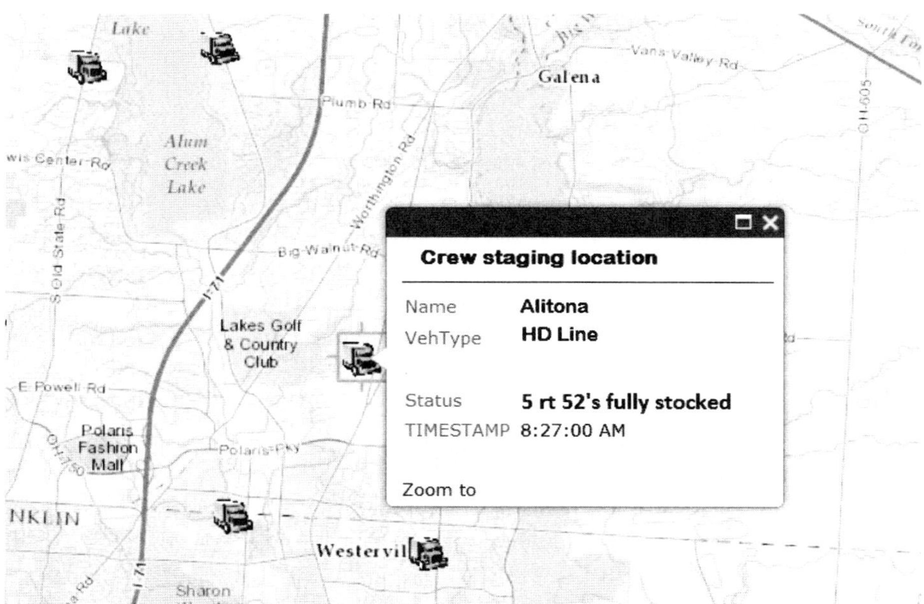

Figure 6.7 GIS map showing locations to stage crews. (Source: Esri et al.)

Response

The response component of emergency management occurs during the event or very shortly thereafter. Depending upon the event, it may be too dangerous for the utility to do anything. Things are often happening too fast. The media, regulators, and public officials are looking for answers. There may be foreign crews showing up during the event. There is a pressure to perform. Customers will often be angry.

Perhaps the most critical thing the utility can do during this response phase is to perform ongoing situational analysis. Here it is critical to gain an ongoing accurate assessment of what's actually going on on the ground. Here the GIS provides a repository for the damage assessment from both the utility and all the agencies involved in the event. Also, the GIS can consume a number of services and map-based data feeds, such as real-time flood data, social media feeds of issues, and any available imagery that may be available that gives utilities a better handle on exactly what they are facing. During this phase, utilities plan their restoration strategies by assessing where they can restore as many customers as possible as quickly as possible. They can plan to isolate those areas hardest hit.

The GIS can consume critical public health and safety information, showing shelters and vital facilities, like water pumping stations, so they will have a rational plan for recovery and restoration. Figure 6.8 shows a GIS map with shelters identified.

The first step during this chaotic period is to gain control over the situation and understand as best as possible the number and location of downed lines and blown fuses.

Recovery

Once the event has passed or it is safe for the restoration effort to begin, the recovery process begins. The challenges during this phase is that there may be difficulty communicating. Customers may have been out of power for several hours and

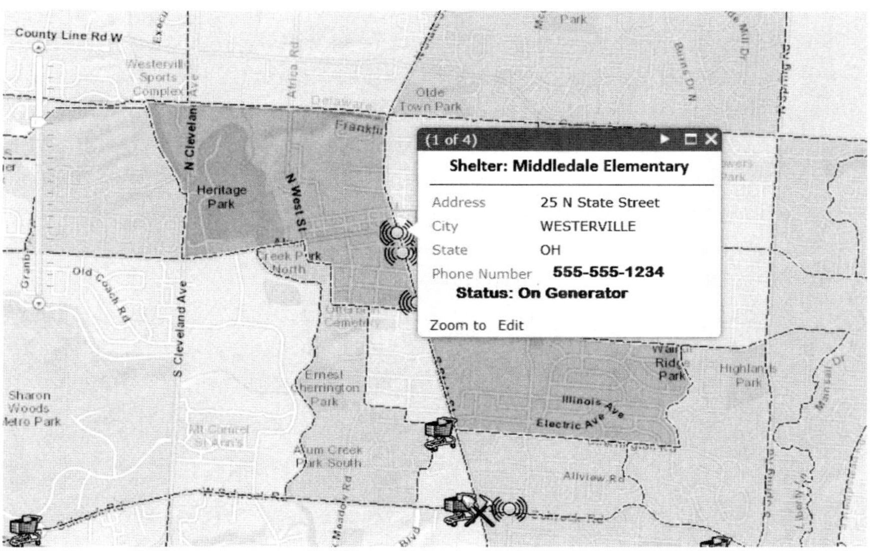

Figure 6.8 Shelters shown on a GIS for utility response. (Source: Esri et al.)

are impatient. The media may be asking when the utility will restore the power. Having built solid plans during the response process will help answer the questions and set proper expectations. During this process, the utility has to be careful to keep track of the material that the crews are installing. Due to the urgency of the situation, the tendency is to get people their power back. However, if the utility fails to document what they replaced during the recovery, they will lose the accuracy of their data.

Utilities need to continuously monitor the situation. They need to fully understand what they have repaired, where the crews are working, and exactly how many customers are still out of power at all times. GIS provides the means to track damage and progress. Utilities use GIS for confirmation of the damage the event has done to their system. Handheld devices are most effective for data gathering and progress reporting. Utilities then adjust their plans based on the continuously monitoring of the situation. Utilities often make the mistake of building GIS mobile applications that are as comprehensive as the office applications used by the GIS professionals. Mobile GIS should be sharply focused on the task at hand. This is especially necessary during the response and recovery phase, where workers in the field are those either from outside the area or are workers pulled from their day jobs.

By having a GIS situational awareness system in place during the recovery phase, utilities can make constant adjustments to their staffing and materials management. Many utilities still use static wall maps and push pins to manage the restoration efforts. Having a solid and agile approach to recover will result in shorter outages, better communication of the situation to outside parties, better cost control, happier customers (if that is ever possible during a major outage), transparency to regulators, and fewer after-the-fact audits.

GIS and Major Emergency Management

GIS is not the only automated tool utilities use to manage a major event, yet it can be a critical element in the response effort. When large proportions of customers are out of power due to catastrophic events, normal processes often get overwhelmed and prove to be inadequate to deal with situation. A well-planned-out approach to a major event using GIS really starts with fully understanding the risk. This may involve elements that are well beyond the control of the utility, but utilities must fully understand them. For example, during the super storm Sandy in the fall of 2012, levies along the Hackensack River apparently breached, flooding the surrounding communities. Utilities need to understand these risk factors and make provisions should such events actually materialize.

Perhaps the most critical information product during the major emergency event is map-based situational awareness. Utilities continuously gain control of a rapidly changing situation by monitoring the situation on the ground and making decisions accordingly. Losing that control increases costs, extends outages times, and results in mistakes and missteps.

The next section deals with the role of GIS during outage management. For any utilities, these smaller outages are just part of the routine of running a utility.

Outage Management

Today most utilities have specialized systems that deal exclusively with outage management (i.e., "routine" outage management). The components and associated outage management of these systems include the following:

- *Information capture modules*—these capture the raw data about outages from customer, police, and fire phone calls, as well as real-time systems like SCADA, and more recently from smart meters.
- *Prediction engines*—these engines take the raw data captured and attempt to group the data into specific events. For example, if all meters reporting an outage are from a single transformer or low-voltage substation, the prediction engine will associate all the meters to a single event. The idea is to locate the actual source of the outage (like a tree falling on a wire) and the location of the disconnecting device in order to disconnect the damaged section from the rest of the system. The output of the prediction engine is a list of jobs that require attention of the field crews.
- *Work management module*—this model takes the list of outage jobs and assigns resources to repair the problems. These modules track the progress of the various outage events from identified, dispatched, and in progress repair to complete.
- *Statistical reporting module*—this module reports on the overall outage management progress, calculating the total number of customers out of the power, the reliability statistics, and often the accumulated cost of the restoration effort.

These specialized systems, called outage management systems (OMSs), are all based directly on a GIS or derive their data from a GIS. Today, utilities are combining the elements of SCADA systems with OMSs into specialized systems called distribution management systems (DMSs). DMSs add functionality to OMSs for real-time data capture, power flow, and real-time switching. As more utilities add sensors to the distribution network and with complete implementation of smart meters, the less the utilities will need to rely on phone calls from customers to determine outages locations. However, the workflows for outage management still remain—the need to determine the location of the outage (whether from calls or from meters), isolate the problem to the fewest possible customers, and to dispatch crews to the right location with the right materials to fix the problem.

Whether a utility uses an OMS or a DMS, the data that each of these systems needs for outage management is the same:

- Accurate representation of the electric distribution system, including a complete inventory of the pieces and parts of the distribution system and their location.
- A logically connected model of the electric network.
- Adequate attribution of the network. This includes essential information about the parts and pieces such as transformer, fuse, and conductor sizes.

- Status of disconnecting devices.
- The condition of the assets. This helps dispatchers determine repair or replacement strategies during restoration.

These requirements define an electric distribution GIS data model at a high level at least for use for outage management. Location, connectivity, attribution, status, and condition are essential for outage management and for most smart grid workflows as well. Not all of this information needs to reside directly in the GIS. The GIS can access attribution, status, and condition from other systems. The key is that the data must be complete and up to date. If, for example, the data used by the OMS or DMS is missing phase designation, then the determination of the exact location of an event will be difficult. Without accurate and up-to-date data, the prediction engine cannot precisely determine the number of actual outage events. If there is a large time gap between when the GIS and the OMS was updated, utilities will have to rely on costly and time-consuming field visits to uncover exactly where the problems are located.

Outage Management Information Products and Workflows

Information products for outage management workflows are as follows:

- Transmission maps (overhead and underground)—these will show access routes and dangerous vegetation.
- Feeder maps (overhead and underground)—these maps show each feeder from the substation;
- Color-coded actual circuit feeder situations—with actual switch positions;
- Outage maps;
- Spot and secondary network maps with outages or problems areas displayed;
- Substation schematics with current operating conditions;
- Display of crew locations;
- Results of system analysis displayed on the map;
- Situational awareness maps—these maps show a combined view of a number of situations that may occur during an outage event, like weather or storm patterns, social network feeds, SCADA feeds, and imagery.

The critical information model for outage management is the complete electric distribution GIS ideally showing the complete transmission network, with HV/HV, HV/MV, MV/MV, and generator substations, the medium-voltage network, the low-voltage substations, and the low-voltage network up to the customer meter or service locaion, plus any equipment, such as a customer-owned generator that could impact safe restoration efforts. The workflow is straightforward—gather all data about where power is out, associate each individual power failure to a cause, determine the location of the cause of the failure, determine the location of the disconnecting device that isolated the power failure, determine the extent of the problem, dispatch repair crews to fix the problem, and document the results.

Specialized outage management systems using the GIS information model directly or as an input to the OMS perform these functions.

In cases of widespread outages, utilities will dispatch scouts to assess the damage and gather the data into the GIS to help manage a longer-term restoration effort. Often these damage assessment applications are quite simple, but of course they are just simple focused GIS applications. Figure 6.5 shows an example of a simple damage assessment application that a scout can use in the field on a simple mobile device like a smart phone or tablet . The information product for this simple app is the map with a series of symbols that show the location of damage to the electrical system. The pop-up would simply display the damage and often a photo of damage site. The damage assessment data then can be transmitted to the corporate GIS where it is combined with the full GIS data to give the dispatchers a better picture of the extent of the restoration effort. This data then becomes available to an overview situational awareness map as shown in Figure 6.9.

GIS Tools for Outage Management

The core workflows of outage management require GIS to perform network tracing. Tracing the network from the customer reports (or smart meter reports) of a power failure up the network determines the most upstream location of the network that feeds the known outages. This trace answers the query of where the most likely upstream location is where power has been lost. Tracing is a core capability of a GIS, provided of course that the network is complete and accurate. Switch positions (open/close) must also be available to the trace process to accurately predict the location of the outage event. Figure 6.10 describes the process for determining outage location.

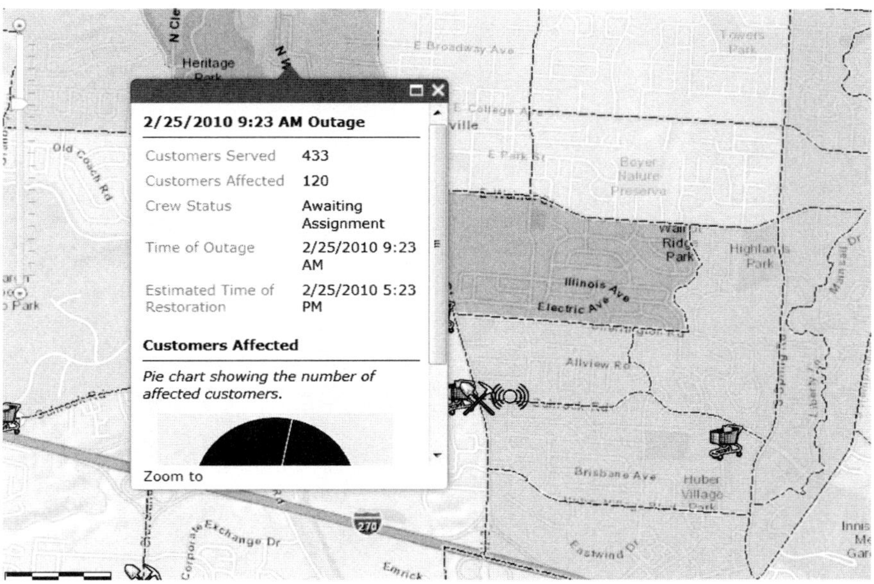

Figure 6.9 GIS situational awareness map during outage events. (Source: Esri et al.)

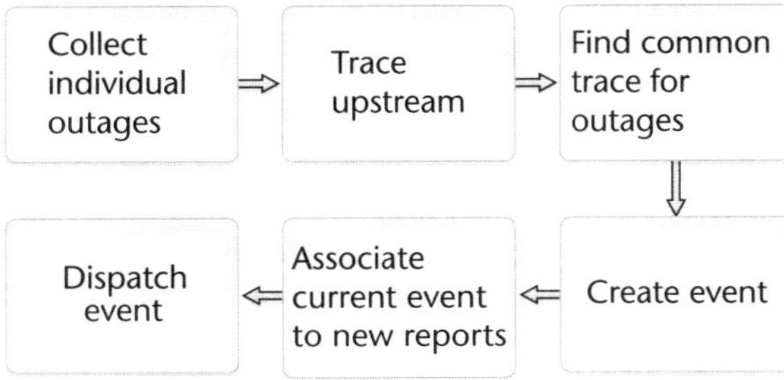

Figure 6.10 Process for determining outage locations.

A variation on the tracing function is to collect network elements used in determining the loading on a section of the network. This application of tracing is needed to determine what the load on a feeder section might be if load were to be transferred from a feeder that is currently out of power to an adjacent circuit. The GIS would trace the feeder, simulate a new feeder section attached to an adjacent feeder section, and then perform a load calculation to determine if the adjacent feeder could in fact handle the added load from the faulted feeder.

Other GIS functions helpful during outage management are statistical functions, such as creating a polygon that adds the total number of point features representing individual outages or point features representing individual damage reports, and then reporting them statistically by a polygon region or shown as a cluster with the total number of outages or damages represented as a point with the values displayed. Figure 6.11 illustrates this concept.

GIS can also provide a way of summarizing outages for reporting externally, such as on a utility's web site. While not showing specific outages, this reporting

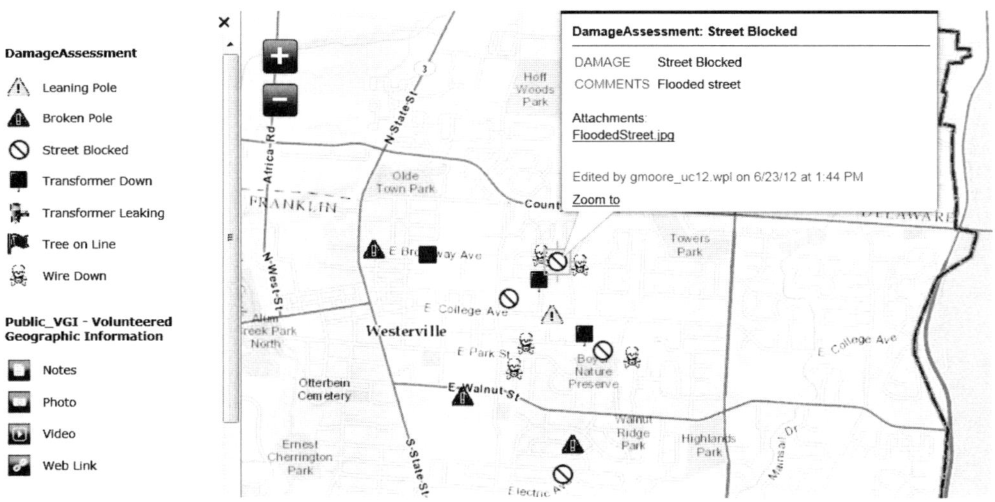

Figure 6.11 Total number of damage assessments shown as a point on the GIS map. (Source: Esri et al.)

can be effective when communicating to a community the regions without power. When reporting outage locations to the public, the utility must balance giving too much detail with not giving enough. Providing detailed feeder information with exact details of the damages would probably not be effective.

The Emergency Management Information Model

The data required for emergency management is captured from both outside sources, such as pre- and post-storm imagery; historic, current, and predicted weather data or from internal systems, like outage management or distribution management systems; SCADA; fleet tracking systems; work management systems; and financial and materials systems. Workflows require the use of all the network information as discussed in Chapter 4. Common information products are varied, but the most important are the damage assessment maps and situational awareness maps. In addition to the information products noted here, these will also be helpful:

- *The location of crews.*
- *The results of system analysis displayed on the map*—as utilities back feed other parts of the system impacted by outages, utilities must be aware of possible overloads, so these maps show the result of load flow analysis using real-time data. It can also simulate the loading in advance of switching to make sure that a switching action doesn't create overloads. Often the analytics are performed by other systems such as OMS, DMS, and network simulation software.
- *Risk profile maps*—utilities can display areas of the system that have higher vulnerability and display that along with the current situation like an approaching storm.
- *Overloaded transformer maps*—if the utility has access to real-time transformer data (from smart meters, for example), they can carefully monitor the overloading of transformers and display the results within the GIS as part of an overall situational awareness.
- *Critical customer locations*—during outage events, utilities need to visualize the location of critical customers. Utilities can publish the locations of these customers from whatever system captures this information, and the GIS can display them along with other relevant data.

Perhaps the most important information product and workflow is the one that no one anticipated. Emergency response is not an exact science, since the extent, location, and cause varies so much. However, GIS is flexible enough to be able to adapt quickly to any number of events, if data about the event is available in a timely fashion. Utilities can create flexible risk models that can adjust to nearly any kind of event.

GIS and Emergency and Operations Management

All emergencies are different. However, utilities tend to handle emergencies (which cause power outages) as either small enough to be able to handle routinely or large enough to declare a major emergency. If the outage event is manageable, the utility has routine processes and procedures in place to handle in a repeatable way. They have systems in place, like outage management systems or distribution management systems, coupled with their customer information system (CIS) that deal with outages in a routine way. During these events, the GIS provides clues to better restoration efforts, a clear situational awareness of the exact progress of the restoration work, and a way to bring disparate data together. Ideally, the OMS or DMS works in close integration with the GIS.

During the abnormal and large events, the OMS and DMS or smart meters don't provide the guidance needed during the early stages of response, since the effort is more focused on logistics, staging, and deployment than understanding which customers are out from which cause. GIS provides the ability to bring together many sources of information, not managed by OMS or DMS, such as road closures, shelters with or without power, flooding, landslides, collapsed buildings, and large areas of devastation.

In all cases, whether the outage event is routine or major, GIS provides tools to help determine vulnerability, help to focus on the consequences of the vulnerability, visualize risk, and provide a means of collecting disparate data, analyzing the data, and providing continuous situational awareness to critical decision makers during a time when customers are most inconvenienced.

Reference

[1] Institute of Electrical and Electronics Engineers (IEEE) Standard 1366-2012, *IEEE Guide for Electric Power Distribution Reliability Indices.*

CHAPTER 7

GIS Enhances the Retail Business

In the old days before utility restructuring, integrated utilities did everything from generating the electricity to rendering a bill to each customer. Today, separate standalone businesses called retailers or competitive suppliers often handle customer care. Yet regardless of who performs the functions of the retail business, the tasks still are the same as in the days before utility restructuring. The distribution business does some of the tasks, while the retailers do others. For example, even though the retailer may render the electricity bill (perhaps even bundled with other services); the distribution company still has to manage customer interaction during a power failure. Or when a customer needs a new service, the distribution company manages that work, not the retailer.

There continues to be a debate globally about which entity—the retailer or the distribution company—should own, install, and manage the electrical meter, smart or otherwise. Both entities need information from the meter. In the U.S., the distribution company universally owns and manages the meters. As a result, the onus is on the distribution company to deliver the meter readings to the various retailers. This chapter will treat all matters of retail and customer care as if the distribution company performs all tasks.

As noted in prior chapters, there are four distinct businesses within the utility: energy supply (production, generation, or energy procurement), wholesale delivery (transmission), retail delivery (distribution), and retail. In each of these businesses, there is a distinct physical demarcation point. The demarcation point for retail is the load side of the electric meter. The distribution business (or retailer) owns the meter and the service to the meter. The customer owns and is responsible for all of the equipment connected to the load side of the meter. There is some inconsistency even with this statement. The service drop is often the responsibility of the customer, even though it is on the supply side of the meter.

The Meter

The meter is at the heart of the retail business. Nearly everything the retail utility business does revolves around the meter. It is the cash register for all four business units. It would seem to make sense that from a GIS facilities inventory viewpoint that GIS captures the location of each of the meters. However, there may be customer equipment between the service entrance or service drop location and the meter. Figure 7.1 shows a simple service with customer equipment and utility equipment shown. Or, there may be a bank of meters in an apartment building basement. Each meter serves a separate apartment. Between the exact location where the service cable enters the basement of the building and the bank of meters, there will be a main service cables, conduit, cable trays, or raceways that run from the where the cable enters the apartment building to the meter bank. There will likely be a splice box where the service cable splits into many cables to feed each meter. All that equipment is customer equipment, including the meter mounting cabinet. If something goes wrong with that equipment, the customer is responsible for the repair or replacement. The only equipment the utility owns is the meter itself.

From a strict asset management perspective, the utility may not want to model assets owned by the customer. There may be some exceptions to this for very large customers, but for most customers the termination point of the GIS should be at the demarcation point of ownership or responsibility between the customer and the utility. That point is usually at the service entrance or service attachment, not at the actual meter location. That means that utilities must maintain a relationship between the actual meter or meters and the service point of the GIS. Sometimes, GIS modelers call this point the premise point, which the utility will often uniquely identify. So the utility creates a one (premise point) to many (meters) relationship within the GIS. In addition, the utility can also maintain a one-to-one relationship between the premise and the service address.

As an aside to this point, utilities should attempt to keep edits to the GIS information to as few as possible, if nothing else to save the labor of constantly editing the GIS. Whenever the utility creates a data structure, it needs to make sure that it doesn't model the data structure in a way that creates unnecessary work for it. A simple example is that some utilities might want to include the customer account number or meter number as an attribute of a point feature called meter. The problem is that meters and customer account numbers can change often, each time a customer moves. However, the only time the premise point changes is when the physical location moves, which is much less frequent and probably would only

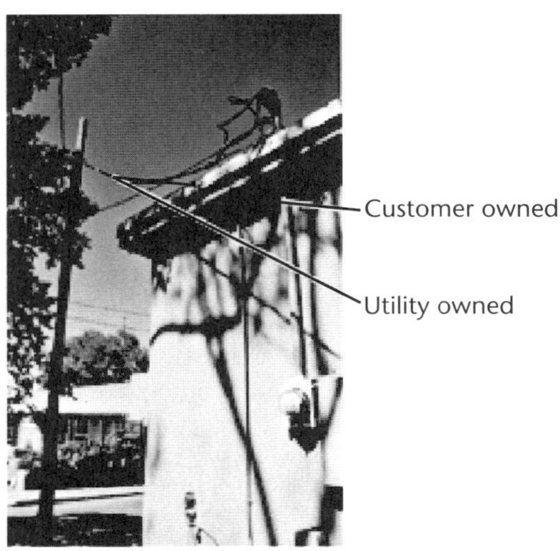

Figure 7.1 Service showing customer equipment and utility equipment. (Photo by Bill Meehan.)

happen if someone was renovating a building. What utilities don't want to happen is that every time a customer moves or they replace a meter, someone has to perform an edit to the GIS. Instead, the utility can create a relationship (outside the GIS) in which the meter (and account number) are related to a fixed premise point.

In this way, whenever the utility replaces, shuts off, or disconnects a meter, the GIS accesses this information easily from other systems, without having to perform an edit on the GIS. In the same way, utilities shouldn't maintain the exact switch position or condition of a fuse directly in the GIS, since they would have to update the GIS every time a switch changes position or a fuse blows. They may want to maintain the desired or scheduled position of a switch, since that is not likely to change much at all. Recall that the GIS is an integration framework, not simply another database to store information. The key information stored in the GIS should be that information uniquely handled by the GIS. Customer information systems, meter data management systems, financial systems, and SCADA better handle real-time status, meter, billing, and customer information. The GIS, however, is the place to bring data from various sources to visualize, analyze, and discover patterns for decision making as well as storing the physical location of all of the utility assets and their relationship to one another.

The Smart Meter

The smart meter will fundamentally change the electric utility business. Utilities that have implemented truly smart meters will no longer have to estimate customer behavior. Utilities will know precisely when every customer loses power, when they have a momentary interruption, what their consumption patterns are (within some time interval), and what the utilities need to do to manage demand. For the entire history of electric companies, utilities have relied solely on customer calls to determine who exactly is out of power. Further, they never knew how much power was flowing through their distribution equipment. For the most part, utilities only knew about power outages when they impacted large portions of the network. For nearly all residential customers, they only knew the average consumption over a large interval, often a month.

The industry refers to the smart meter, its associated communication network, and head end gear as advanced metering infrastructure (AMI). Industry leaders crafted this term to differentiate AMI from automated meter reading (AMR). While AMR reads consumption in a somewhat automated way using various techniques, it does not provide the complete infrastructure for two-way communication with the meter. In addition to the utility communicating with the customer meter, the AMI infrastructure allows the utility to communicate to devices beyond the meter. So the utility could, if allowed to, disconnect individual loads inside the customers' facilities. When fully deployed, utilities could control demand at a very granular level during high peak demand days or even adjust demand for dynamic pricing. However, AMI still does not have knowledge of how the utility delivers the electricity. It doesn't know which feeder is supplying power to the customers. Further, AMI has no knowledge of causes of customer behavior. AMI reports only outcomes. The GIS, together with AMI, SCADA, generation control, and DMS, provide utilities with a complete picture of the entire supply chain of the utility delivery in nearly real time. The GIS provides the network infrastructure information, plus the spatial analytic capabilities to tie all the pieces together from the outcomes to the causes.

AMI Network

There are three major components of AMI: the collection of smart meters, the meter data management (MDM) system, and the communication infrastructure or AMI network. The purpose of the MDM is to consume the data from the meters via the network and process the data for use by other systems, like billing or outage management systems.

The AMI network consists of all the devices and equipment between the smart meters and the MDM. Utilities have a number of options regarding the AMI network. Some utilities have decided to build, own, and operate this network themselves. Others decided to use commercial networks, while still others use a combination of commercial and private networks, in which one backs up the other. Most utilities have some form of telecommunications network already. Many will leverage the investment in their existing network. Figure 7.2 illustrates the components of AMI and their major functions.

In all these cases, the location of the equipment is critical for operation of the AMI network. Modeling any telecommunications network in GIS is beyond the scope of this book and is as complicated, if not more complicated, than an electrical GIS. It merits an entire book of its own. The option the utility used for their AMI network (private, commercial, or hybrid) determines how much control they have over the physical communications network itself. Just because the utility does not operate the AMI network, doesn't mean that it cannot access the information about the AMI network. If the utility utilizes a communication network they do not own, they can craft an information-sharing agreement with the carrier. For example, the carrier could publish data as a GIS web service to the utility of all their critical facilities that impact the AMI network. The web service doesn't have to have the detail that the carrier owns, but it does have to have enough detail to be able to correlate the AMI network components with the utility's infrastructure.

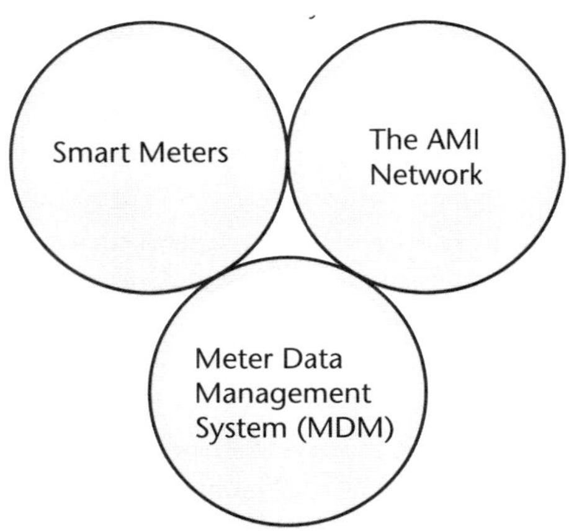

Figure 7.2 The components of AMI.

For example, if an errant driver knocks down a pole that has the electric equipment and a critical cell relay for the AMI network, the utility needs to know that communication from a group of meters may be impacted.

If, on the other hand, the utility has built out a completely private network, then they need to create a telecommunications GIS in much the same way that a commercial carrier would. The electric and the telecommunications GIS can share layers about structures (such as poles and conduits) and base maps, but they can manage the facilities separately. However, both the electric and the telecommunications GIS can easily share information with each other. They can both participate in an output products and spatial analysis models. As noted, they certainly can share base map information from the same source.

Meter Data Management (MDM) Systems

The MDM processes the raw data that it receives from the smart meters (and other meters as well), analyzes network performance, reports on meters not reporting, and creates meter-to-cash processes in preparation for billing. The vision of smart grid is to use the MDM-processed data to give customers and utilities information they can use to manage consumption.

One of the original concepts of smart grid was to reduce greenhouse gas production. Figure 7.3 illustrates the principles of smart grid as outlined from the U.S. perspective in the law that the U.S. Congress passed in 2007. One way to do that is to manage demand. This notion called demand response started with providing consumers ready access to their consumption information, well before they received their monthly bill. Given that information, proponents of the smart grid argued, consumers could alter their consumption behavior well in advance of receiving their bill. The other way to impact demand was for utilities to actually take direct control of a consumer's consumption. This would happen by actually disconnecting devices during times of heavy demand. The other concept was to alter pricing of retail electricity depending on the wholesale pricing. Utilities call this dynamic pricing. Dynamic pricing is an extension of the time of use pricing. Time of use pricing assumes that the wholesale cost of electricity is lower during off-peak times (from, say, 6 PM to 6 AM) than during peak times. Dynamic pricing extends the concept to actually follow the wholesale cost of electricity, so, theoretically, pricing of electricity could vary throughout the day depending on the cost of power produced.

MDM provides the means of analyzing the raw data, processing the data, and providing utilities the knowledge of how to best utilize demand response.

Role of GIS in AMI

Clearly AMI will improve the billing process. It will provide an automated means of disconnecting and reconnecting service to customers. Without AMI, utilities either have to visit the site of a meter to disconnect or reconnect or leave the service connected. Neither situation is ideal. Remote disconnection avoids what utilities call a truck roll, having a truck drive to a location to disconnect or reconnect a meter,

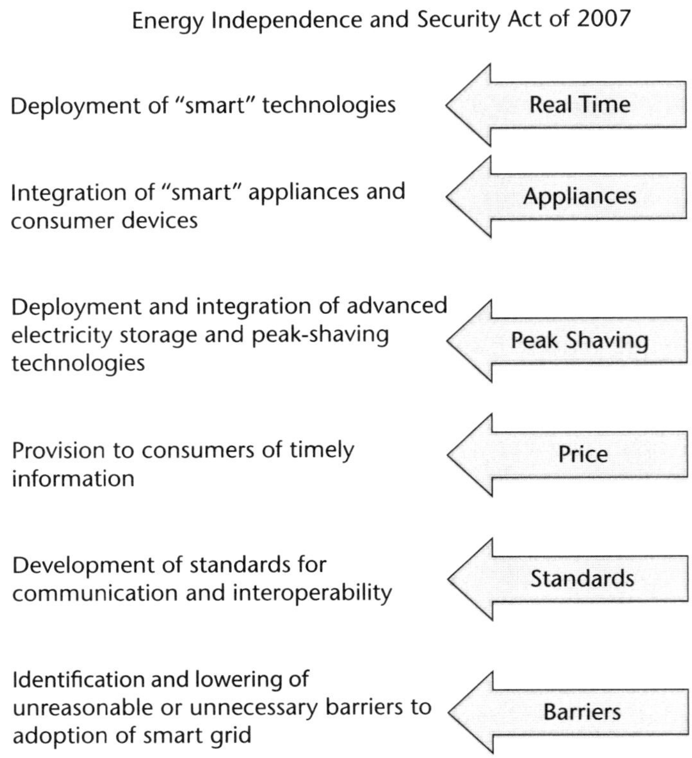

Figure 7.3 The smart grid as envisioned by the U.S. Congress.

which is expensive. Leaving the meter connected often results in unauthorized use of the electricity. Utilities can immediately get a final read of the meter for customers vacating or an initial read when a new customer moves in. In the unfortunate cases of customers not paying their bills, the utility can shut off a customer for nonpayment remotely and reinstate power as soon as the bill is paid.

Utilities can enrich the data from AMI by integrating the consumption data with GIS. In these cases, utilities can see patterns of consumption. Then they can see where potential demand problems may be developing well before situations get critical. They will be able to see where to target marketing programs for demand response. The idea would be to discourage usage in those areas where electrical capability is limited. They could encourage customers to participate in remote disconnection of appliances in those areas where it makes the most sense. Figure 7.4 shows smart meter information on a GIS map.

In effect, the GIS together with the MDM provide a situational awareness of exactly what the consumption patterns are much faster than ever before. Generally, they can view consumption accurately the next day for the prior day.

Customer Care Business Processes and Drivers

Customer care involves a number of processes as shown in Figure 7.5.

These processes are as follows:

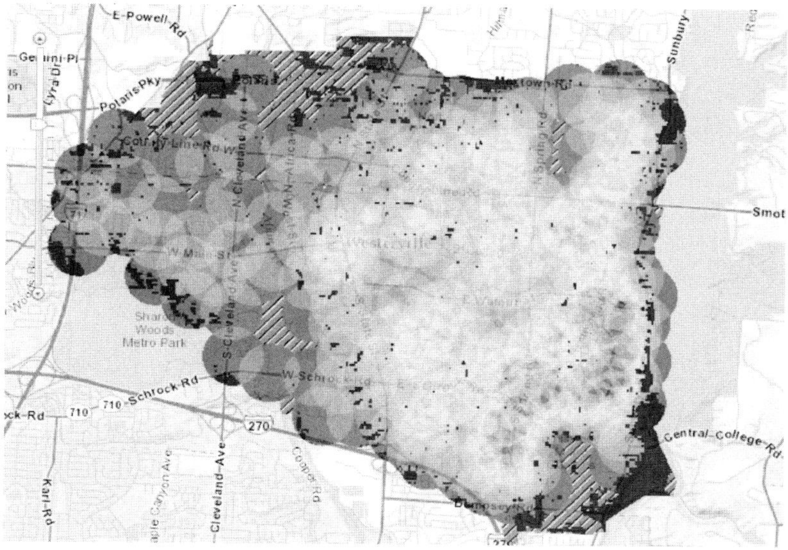

Figure 7.4 GIS web map shows smart meter consumption data. (Source: Esri et al.)

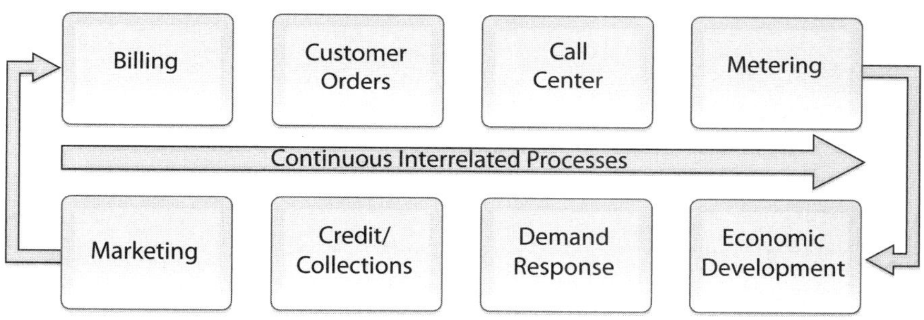

Figure 7.5 Customer care business processes.

- *New service connections*—creating work orders for new facilities to supply power to new customers;
- *Billing*—processing the consumption data and rendering the bill to the end use customer;
- *Credit and collections*—managing delinquent accounts with the intent of minimizing bad debt write offs;
- *Metering, smart or otherwise*—installing and removing meters, assigning meters to customers, calibrating, testing, buying, repairing, and disposing of old meters;
- *Call center*—managing staff who answer customer phone inquires; these calls may be about outages, billing, or connecting, disconnecting or installing new service connections;

- *Marketing*—encouraging customer behavior to participate in a specific program, such as conservation, load management, or weatherization;
- *Economic development*—encouraging business to locate to areas that benefit the electric company, specifically to those areas with spare capability;
- *Demand side management or demand response*—managing programs to reduce consumption, such as relamping factories with energy efficient lamps;
- *Account management*—assigning staff to deal with large customers.

GIS and the New Customer Connect Process

Chapter 5 dealt with the work order process. In this process, the utility receives a request to alter the distribution system in some way. The request may come from the planning or asset management department for an upgrade to a feeder or from the city to relocate wires due to a street widening project or from a reliability engineer to break up a feeder into smaller sections or add a new automated sectionalizing switch. Perhaps the most common request comes from a prospective customer for new service. This request ranges from a new overhead service from an existing overhead distribution line to a feeder supplying a new shopping mall or housing subdivision.

While the work order process is the same as outlined in Chapter 5, the nuances for customer care are a bit different, since the requestor of the change to the infrastructure is a customer. Processes need to be in place to satisfy the customer as well as to assure that company implements the proper metering and starts the revenue flowing correctly. The major difference between a work order to upgrade a feeder to a new customer connection is that the new customer connection request happens randomly and unexpectedly. This means that the utility needs to have agile processes in place to move the request along to get the project completed when the customer needs the service. The utility could receive a flurry of new connections at any time.

The new customer connect process involves a number of steps; the very first is negotiating an in-service date with the customer. The second is negotiating a price for the new service. Utilities pride themselves on providing good customer service. One common metric of customer service is meeting customer in-service dates. Once the utility and the customer agree on an in-service date, the customer in-service clock starts clicking. If the utility fails to supply power to the customer on that date, the company has an unhappy customer and the failure detracts from meeting a key performance metric. Delaying electric service can have a direct revenue impact on revenues as well.

The utility must make a rapid assessment of the request to determine whether the present infrastructure has adequate capacity to handle the new load. Whether or not it does impacts both the cost and the schedule of the project. It also has to determine whether the new construction requires governmental permits, such as easements, street opening permits, or grants of location for new equipment. Finally, the utility must determine if it needs any special metering equipment or if it needs to extend or upgrade the AMI system.

Role of GIS in the New Customer Process

GIS can help automate this process and provide visibility into how well the utility is performing. Figure 7.6 shows a GIS map with customer work orders identified. The GIS will have the key data to determine whether the utility needs to upgrade the infrastructure (both electric power and smart grid). The faster this is done, the sooner the utility and the customer can determine both the costs and the schedule of completing the new service.

The traditional way utilities handle this process often goes like this:

- The electrical contractor contacts the municipal agency requesting a wiring permit for the electrical work they need to do, including the new electrical service. The contractor will likely have to explain all the details of the project to the agency.
- The electrical contractor, with wiring permit in hand, either meets with a utility representative or calls the utility with a request for a new service. The contractor repeats the same information to the utility contact person that he or she told the municipal agency. The contractor requests an in service date. The utility accepts the date but does not confirm it. The utility then explains about possible costs, but due to a lack of information cannot provide exact costs, nor can they collect any fees in advance of the work.
- The utility representative places the request in a work management queue. The contractor does not have any guarantee on price or schedule.
- Depending on the workload, the utility engineers or designers review the request to see if the present infrastructure can handle the new service. Of course, this has to be done in relation to other requests that may be in the same area. To adequately do this, the engineers must request loading in-

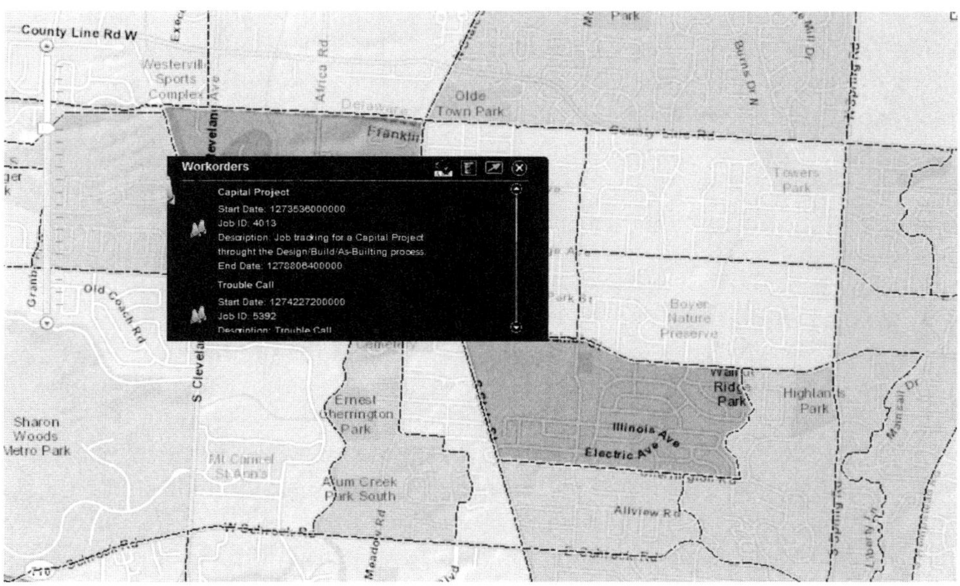

Figure 7.6 GIS map showing customer work orders. (Source: Esri et al.)

formation from the billing department. They then must perform a network calculation to determine if they need to upgrade the infrastructure.

- If the infrastructure is adequate, then the engineers perform a preliminary design and determine the cost of the service. They then need to schedule the work based on the current workload and the available resources. If they determine that they need a new transformer or they have to upgrade a feeder, for example, the project becomes much more complicated, costly, and uncertain.
- Once the designers complete the design, they determine if they will need to obtain any permits from the government. Of course, they won't be able to determine if they need permits until they understand the full scope of the work. For example, if the utility needs to install a new transformer because the new load will exceed the existing transformer, that work may involve permits.
- Once the utility knows the scope of the work, they can they predict when the work will be completed and what the costs will be.
- Once they receive the notice to proceed from the customer, the work proceeds like any other new construction project.

The problem with this workflow is that this assessment process can take quite some time, delaying the project and introducing a degree of uncertainty for the customer. One way to avoid this delay is to provide a GIS portal for customers (contractors) that allow them to enter their load data and any special requirements directly into an application that is either GIS-based or has GIS functionality. The portal can quickly determine if the utility will need to provide additional infrastructure or not. If the service request is routine, meaning that it will not require anything special, the contractor can directly initiate the work order, pay the connection fee, and receive an automatically created in-service date. For special situations, the contractor will receive a notice that the request requires additional engineering. This process eliminates routine projects from clogging up engineering queues.

The utility could create a simple arrangement with the municipality to automatically transmit wiring permit data to the utility. Having this arrangement in place, the utility would know much of the information it needs for the project, well in advance of the time the contractor actually contacts the utility. It's possible that those more complex tasks could be pre-engineered as well. This self-service idea shortens the time between when the contractor initiates the wiring permit and the determination of the in-service date. In fact, the contractor could communicate with the utility directly at the job site in the field using a mobile device.

For very simple connections, the utility could automate the entire process right up to construction. When the utility receives a request for service, the self-service GIS-based application validates that it is a simple request, initiates the work order, receives payment by credit card, and submits the request directly to the field. It is possible that the utility could connect the customer within hours of the request. To do this, the GIS, the work management system, and perhaps a network analysis system (depending on the complexity) needs to communicate and integrate. The GIS application also has to have up-to-date loading information available from the MDM or customer information system has to be able to determine the current loading.

Most requests for new service are simple. Yet many utilities take days or weeks just to get the requests into the field for construction. The delays are a result of handoffs from one group to another—from the work order group to the engineering group, to the permit group, to the billing group, to the meter group, and back to the work order group, with approvals and wait times in between. Since the GIS accesses nearly all the information needed to satisfy the request (along with integration with other corporate systems), utilities can eliminate much of this delay.

GIS and Billing, Credit, and Collections

Billing is the processing of consumption information, applying the rates to the raw metering data, and rendering a bill to a customer. Historically, the vertically integrated utility performed this process. For most of the utility's customers, this process is routine. The utility sends the bill, and the customer pays. However, there are certain customers who do not pay or do not pay on time. These customers create the most work for the billing department, and the lack of payment negatively impacts the company's revenues. Every year, utilities give up on collecting money from delinquent customers and simply write off the receivable. The accumulated write-offs are referred to as bad debt. Bad debt for some utilities can be high. For some utilities, the dollar amounts can be in the multimillion-dollar range. Utilities may decide certain approaches prior to writing this debt off. They can actually visit with customers (knocking on doors), call, email, or threaten disconnection. All of these approaches are expensive.

The utility can use GIS to help them cluster delinquent customers to optimize the visits. They can map the distribution of customers based on the time they are late. They can analyze the patterns of payment for certain customers in certain areas.

Utilities must be very careful not to profile groups of customers. However, by knowing payment patterns and demographics, they can fine-tune their collection activity to optimize collection success while minimizing disrupting disconnections. Figure 7.7 shows how GIS maps can illustrate demographic patterns.

Utilities can also market prepayment plans to certain customers. For example, a utility might wish to market prepayment to areas with a high percentage of college student populations. Prepayment avoids the collection activity completely. Utilities can map historic areas of their service territory where people move frequently. This gives them the focus for their prepayment marketing activity.

GIS and Metering

Metering, smart or otherwise, involves the entire process of the management of the customer meter. When a customer requests a new service, the utility must provide, install, and commission a meter for that customer. The utility must set up a new account, and the utility must notify the GIS that they have created a new service

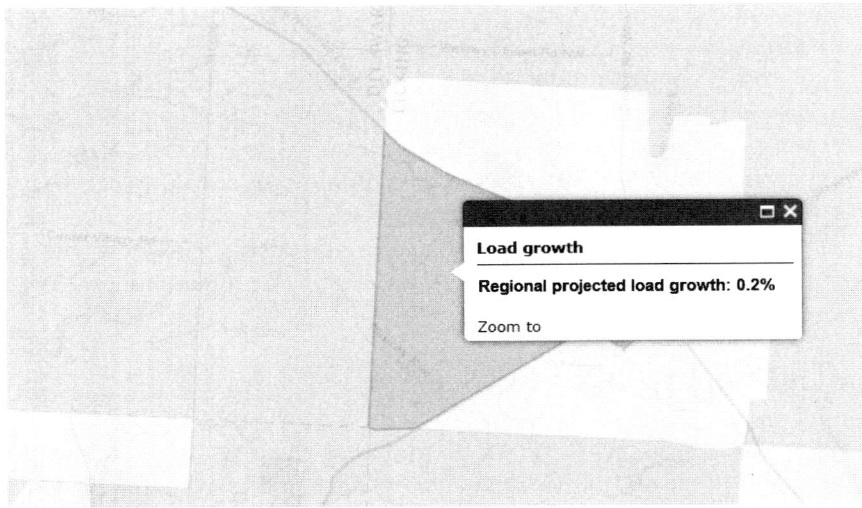

Figure 7.7 GIS web map showing demographic zones. (Source: Esri et al.)

location. The process has to be tightly coordinated with the new customer connect process noted earlier.

Meters fail and get old. In cases where the utility migrates from old meters to new smart meters, they have to schedule the work. The logistics can be daunting.

Since the GIS maintains the service location, utilities can use it to manage the logistics of meter repair, meter testing, and replacement. Chapter 8 will go into more detail on routing and field force management. Utilities use GIS as a situational awareness tool to visualize what is going on specifically regarding meter issues. Knowing the clusters of smart meters not reporting, the utility can determine if meters are failing or, more likely, coverage is lacking, or perhaps the network has a problem. Figure 7.8 shows how GIS is used to identify problems with smart meters.

Customers will complain about high bills. Most often these complaints are a result of high consumption. However, if the GIS notes a pattern of high bill complaints clustered in a certain area, they may be able to determine that there is something going wrong in the area. GIS provides the tools to give the entire metering process a spatial context. It can help optimize meter testing and schedule new meter installations routes. With an entirely new replacement of meters, the utility can use GIS to report on progress, program in the best paths for replacement, and create reports for senior management that shows the progress of the deployment.

For those utilities that have not deployed smart meters, they can utilize GIS to identify tough-to-read meters. Utilities do not like to estimate bills. They like to read each meter on a scheduled basis, so they can be sure that the bill that they send to the customer accurately reflects the customer's consumption. When they cannot gain access to the meter to read, they resort to sending an estimated bill until they can gain access to the meter. If they have underestimated a customer bill for several billing periods, when they finally get an accurate reading, the customer could get a very large bill. This situation creates problems for the call center, the billing department, and, if the situation is severe enough, a story about this ends up in the newspapers. Estimated meter reads can be very expensive for the utility. So utilities can

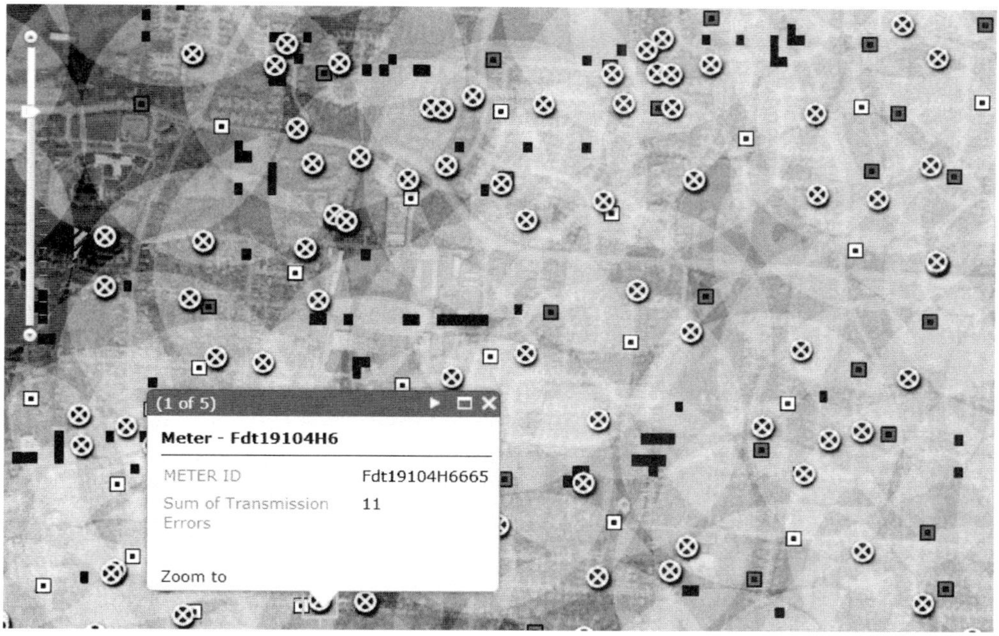

Figure 7.8 GIS web map showing problems with smart meters. (Source: Esri et al.)

map where accessibly problems exist and help utilities to target areas for a limited deployment of smart meters to those areas with the greatest accessibility problems.

Today, many of the countries in the world and states in the U.S. have unbundled the retail and billing processes. In the U.K., for example, there are seven distribution companies and nearly twenty energy supply companies. The distribution companies and the energy supply companies need to share the consumption information. In Texas, for example, the distribution companies own and operate the meters but deliver the data to a central clearing house where the retailers access the data.

GIS can be used as a collaboration tool for the distribution companies to be able to identify exactly which retailers service which customers. When things do go wrong, the retailers will be better informed when customers contact them.

The Call Center

The call center consists of a number of employees (with access to the utility's corporate information systems) who wait for customer calls. The call center employees have access to data about the customers. Often utilities will have overflow call centers in other parts of the world to deal with times when the volume of calls exceeds the capacity of the utility call center. Some utilities outsource the entire call center. For many customers, the interaction between themselves and the call center staff is their only touch point with the utility. This means that the responsibility for customer satisfaction depends heavily on how well the employees handle the calls from customers. Call centers handle the following:

- *Outage calls*—though most utilities now handle outages via an interactive voice response (IVR) system;
- *Exit or entering calls*—customers who plan to move contact the call center to cancel or move their service to a new address;
- *New customer calls*— a customer who moves into an existing facility that has an existing service already installed;
- *Billing inquiries*—calls from customers complaining about a high bill or want information about their bill; often a customer will complain about an outage due to a disconnection for nonpayment;
- *New customer connections*—calls from a customer representative (often an electrical contractor) requesting that the utility install a new service (and meter); see earlier discussion in this chapter;
- *Street light outages*—calls from customers or others reporting a street light not operational;
- *General complaints*—these are calls from customers and others complaining about nonelectrical situations like utility trucks parked on the street blocking passage or tree trimming crews cutting down trees.

The key to successful resolution of customer calls at the call center is to make sure that the call center representatives have timely and complete information available to them. Most of the calls they receive have some kind of spatial component to them. Certainly it is helpful for the call center representative to know where the customer is calling from based on access to the GIS.

Street lights are not only for public convenience; they provide a protection against crime in neighborhoods. Often utilities are measured on their response to street light outages. In some cases, street lights are owned by a number of entities, so the utility needs to know which lights are theirs. When a customer reports a street light out, the call center person needs to be able to visualize on the GIS exactly which light the customer is talking about. Asking the customer to identify the street light by some pole number attached to the pole is not effective, since the customer is likely calling at night and the street light is out, so identification will be difficult. Instead, the call center representative can see on the GIS which street light the customer is reporting. Utilities can use the GIS to automate this process as well. Customers can call an online GIS map (on their smart phones or tablets) showing street light symbols. The customer can simply click on the symbol and report the outage. Then, if a customer does call the call center, the representatives have immediate access to the status of the street light problem and could confirm that someone else has reported the problem.

A GIS-based street light self-service system routes the report to the street light crew. Having these GIS displays available to call center staff enhances employee morale since employees will have the right tools to satisfy customer requests at their fingertips. Not having good tools can frustrate staff, since they will be dealing with frustrated customers. Customers appreciate accurate answers to their questions. Giving employees timely and complete information improves cutomer satisfaction.

During outage events, the call center representatives need to know the extent of an outage, where the crews are located, and the estimated restoration time. This will help customers and business owners impacted by the outage. Business owners will have a better handle on whether to close, stay open, or when to let employees go home.

As noted, utilities use GIS as a situational awareness tool to give to call center staff the locations of outages and of crews. In addition, many utilities publish an online outage visualization tool, driven by GIS. Often, if power out is widespread and customers don't have access to information, the call center is the only source of information for them. Many utilities use social media to help get more information about what is going on in their service territory, as noted in Figure 7.9.

Certain jurisdictions allow utilities to disconnect customers when they are delinquent on their bill. It's important for the call center staff to understand the timing of this event. If the customer pays the bill, but the utility has a crew on the way to disconnect the customer, the call center person will need to have visibility into the progress of the disconnection process. They will need to know the location of the crew assigned to disconnect. If they have not disconnected the customer they will need to notify the crew that the bill has been paid. This avoids the problem of a crew disconnecting a customer after the bill has been paid, as well as the added cost of returning to reconnect service to the customer. Using the GIS for situational awareness into the status of disconnections is critical. This will be especially valuable for a utility with smart meters.

Geomarketing for the Retail Business

Geomarketing is the practice of using location to fine-tune messaging to targeted segments for the purpose of increasing the attention of a particular segment. For example, if a utility wants to market a prepay program, they need to segment the customer base into those areas where prepay has a high degree of favorability. If, for example, the utility wants to market green power, then they would target the segment of their customer base that they know would be willing to pay more for green power. These two programs would appeal to two very different types of customers.

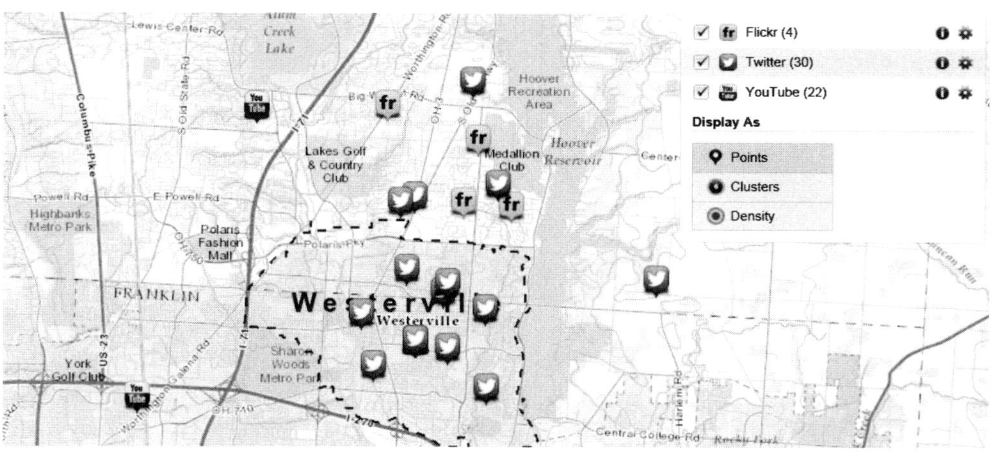

Figure 7.9 Use of social media incorporated into GIS. (Source: Esri et al.)

Those types of customers, however, do not tend to live in the same neighborhoods. The notion of geomarketing is to match the demand—the product or service being promoted by the utility to the most likely customers who live or work within certain areas of the service territory.

The criteria for these areas tend to be based on demographic studies, including the average age of customer, their average income, their buying habits, their hobbies, and their interests. Utilities (and retailers) use this demographic segmenting to effectively market a particular program to the audience.

For many years, utilities used the term *ratepayer* to describe the consumer of electricity. Only recently have utilities abandoned using ratepayer and adopted the more appropriate term: the customer. The only real differentiation utilities performed on their ratepayers was, not surprisingly, by their rate class. So when a utility did marketing, their only real source of data was their customer information system (CIS). The CIS takes the rate and the consumption to create the bill. Utilities have adapted CISs to capture customer interactions, but utility CISs are often not designed to help utilities truly understand customer behavior beyond their payment, complaint, and maybe outage and inquiry history.

What utilities have discovered is that for years the consumer retail business has been segmenting customer behavior so they can match their offerings to the segmented and targeted customers who are most likely to buy their product. What retail businesses have learned is that location really matters when attempting to understand customer behavior and their propensity or affinity to certain products and services. Location helps businesses understand demographic, psychographic, purchasing, and spending characteristics.

The first step in target marketing is knowing where the customers are. A utility knows this exactly since all their "ratepayers" are located within a defined service territory. Knowing the concentration of their segmented customers saves utilities from canvassing their entire service territory to uncover that only a certain demographic is interested in a specific product offering. In other words, using GIS, utilities can discover their customers' lifestyle characteristics, buying habits, how far they are willing to travel, and what they are buying. They will be able to see where people of similar characteristics are clustered. Then they can target market directly to them.

Utilities that operate overhead electric lines will most likely have to perform tree trimming. Some utilities spend enormous sums on tree trimming alone. Yet, despite the obvious benefit of tree trimming, there is a segment of the population that abhors any tree trimming done by utilities. Utilities continue to trim trees in areas where opposition is high, without regard to marketing to that segment of the population. Then the utilities are surprised when they get negative press about their "callous trimming of the beautiful palm trees." Utilities could segment their customer base, market a healthy-tree program, and invite customers to learn about the advantages of trimming trees. While this isn't marketing a product or service directly, it uses the principles of target marketing to achieve the goal of matching customer needs to business offerings.

The next step is to segment the customer base by lifestyle characteristics. The result of this analysis is a series of polygons, with each polygon sharing a dominant set of lifestyle characteristics, like enjoying golf, dining out frequently, and owning luxury cars. Other lifestyle profiles are those people who might be likely to camp out, buy from health stores, wear jeans, and ride bicycles. With this data, utilities can learn much about what their customers are like, not just what rate class they belong to.

The third step is understanding the targeted customer segment's needs and desires. This matches who the customer is to what the customer is likely to buy or respond to. If, for example, the utility is targeting customers for a time-of-use rate structure, they would target people who are likely to change their consumption behavior.

The last step in geomarketing is to analyze the offering and match the offering to the segment most likely to benefit or buy into the offering. Customer segmentation based on geographic and lifestyle characteristics plays an important role in determining what segment to target. This assures that utilities are not spending marketing resources in areas unlikely to produce results. So as utilities move from thinking of customers by their rate class to their lifestyle characteristics, they can better match their offerings or even their core services to the right customers.

Today, demographic, psychographic, spending patterns, and customer segmentation data is widely available over the web. Utilities can simply mash up lifestyle data with their customer information data and easily perform spatial and location analytics to properly segment customers.

While many utilities participate in social media, few use social media tools to help in their marketing. In addition to the customer segmentation, utilities can mine social media to determine what customers are thinking (and tweeting) and where they are located. Utilities then can perform spatial analysis on demographic data and social media data to uncover how customers may react to a specific program well before they even launch the campaign.

Economic Development

Utilities can perform a different kind of marketing that attracts new business into their service territory. Unlike target marketing, in which the targets are existing customers, utilities will market to prospective customers. What a utility is marketing is not a product or service, but an actual pricing structure or discount on electric prices if a new business locates within their service territory.

The reason a utility would do this is to increase the utilization of their infrastructure, particularly in areas where the infrastructure is lightly loaded. If a substation or feeder is loaded to only 50 percent of its rated capacity, then a utility could generate additional revenue without having to invest in infrastructure if they can encourage a business to relocate to where the electrical capacity is plentiful. Here the utility can use the GIS to answer the question: find vacant parcels or buildings of a certain size connected to the targeted substation or feeder. They can add other criteria, such as finding a parcel near a freeway off ramp or near a commercial airport or close to rail transportation.

Utilities can create a GIS spatial analysis model that allows the utility to vary the requirements (e.g., parcel size, available parking, costs, and distances). Again, like in target marketing, the GIS matches a customer need to the utility's offering. For target marking, location is critical.

GIS and Customer Care

The customer care organization in a utility is essentially the retail business. The installing, testing, management, and reading of the meters are fundamental to billing, and so this is one of the most critical aspects of the customer care processes. GIS allows utilities to locate where problems are, provide locational data as to how to respond, and provide logistics support for the meter process. As more utilities implement smart meters and AMI, they will want to leverage the consumption data and map it and analyze the patterns. This will provide input into their demand response programs.

The new customer connection work order process for new business is a key business activity. Customers will rate utilities on how fast and responsive they are to their requests. Utilities need to be agile since they don't know when these requests will happen. They can use GIS to help manage this process and even automate many of the more routine requests.

Location matters for billing and collections. Utilities continue to struggle with collection activity. Utilities can use GIS to locate areas where delinquent accounts are most prevalent and help them craft programs to mitigate collection activities by matching customer behavior with collection activity.

The call center staff often provides the only interaction between a customer and the utility. Giving them the right tools and the visibility to what is actually going on within the service territory goes a long way toward improving employee morale, since an uninformed employee in a call center is apt to get frustrated quickly. Giving call center staff tools such as being able to locate a burnt out street light, knowing where repairs crews are working, and seeing where customers have been disconnected for nonpayment helps to provide superior customer service.

Finally, GIS provides a fundamental tool for marketing activities, from target marketing for products and services to influencing behavior for operational issues or new construction issues to economic development activities. Utilities will need more than rate class to differentiate their customers if they plan to fine-tune their demand response and other marketing activities. Since customers are distributed throughout the service territory, location matters. That's why GIS enhances the retail business for utilities.

CHAPTER 8

GIS and Shared Support Services

Electric utilities have used GIS (or at least digital mapping) for several decades. However, the primary use has been for documenting the location of facilities and, of course, creating operating maps. So the users have tended to be engineering and operations employees. As noted throughout this book, even in those organizations, GIS has not played a prominent role in spatial analysis or location analytics. Neither has GIS played a role as a framework for integration of many kinds of data and systems. The use of geospatial technologies to view information coming from real-time systems (like SCADA, DMS, weather, and traffic systems) semi-real-time systems from meter data management (MDM) systems, predictive engines like lightning and flooding prediction services, in additon to identifying the location of work orders and crews have not been widely adopted, although this seems to be changing.

So while plenty of opportunities for GIS in engineering and operation are still untapped, there are many opportunities in areas that rarely utilize geospatial technology. Yet much of what an electric company does, even outside of operational areas, involves the dimension of location. For example, where are utilities spending money, having environmental issues, having customer satisfaction problems, or having higher than average motor vehicle accidents? Where should utilities locate their warehouses, storage yards, and spare parts? What land do they own? What land use licenses are about to expire? Where do employees live? This chapter details how GIS can support the many other departments within a utility and how the information from operations and engineering can influence decisions in the support areas and the other way around. A true enterprise GIS would serve a maintenance worker repairing a coal conveyor in a power plant to the CEO watching spending distributed throughout the distribution system. Figures 8.1, 8.2, and 8.3 show the processes associated with shared services activities in a typical electric utility company.

Managing Land Information in GIS

A GIS is of no value without information about the land. Utilities select sites for power plants. Transmission lines deal daily with issues of the right of way.

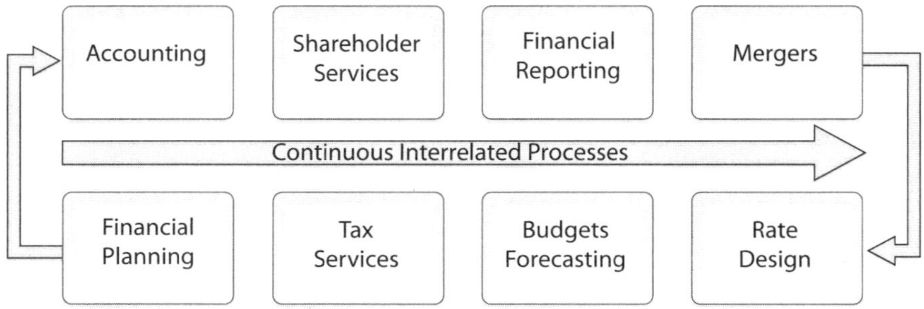

Figure 8.1 Financial shared services processes.

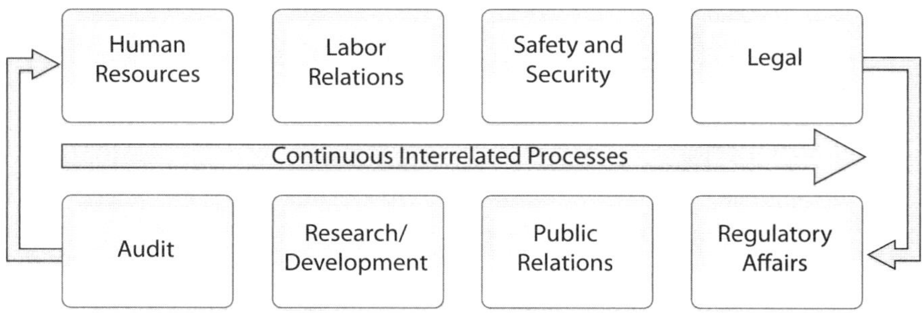

Figure 8.2 Corporate services shared services processes.

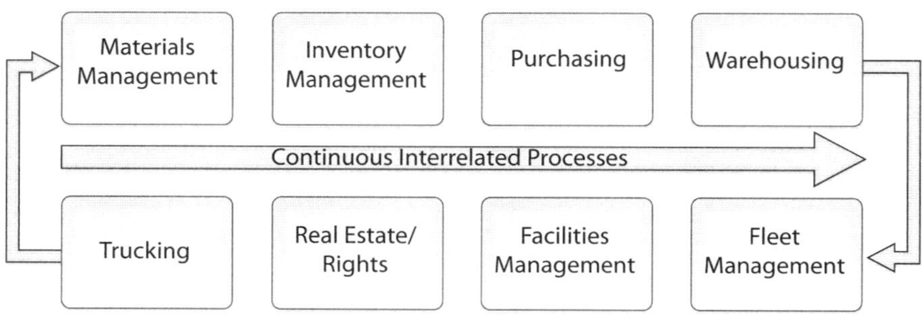

Figure 8.3 Supply chain and logistics processes.

Distribution utility trucks drive up and down city streets. Retailers need to know where customers live. Clearly, the GIS has to have a model of the land features. Most utilities deal with land information in several organizations, namely, the real estate, the surveying, and the rights and permits groups.

GIS for Real Estate

The real estate group manages land holdings of the utility: buys and sells property, pays real estate taxes, and perhaps maps the holdings. The survey group creates the property attributes of holdings, easements, and licenses and records the legal descriptions of all property the company has interest in. They may also make maps of the utility's various land holdings and easements. The rights and permits group requests various grants of location and permits from the government to be able to install electrical facilities in the public ways. They, too, will create maps showing where the utility will need rights and permits. They may also create maps that show the result of the grant from the government. They will record any time limits on any grant, right, or license. They will record any and all conditions associated with an easement.

The Land Base

Regardless of who actually manages these functions, the end result is information about land ownership in the form of maps. Each of these groups act independently—the data they create is often contained within their own group and is hard to share. Even if they use GIS technology, these maps and systems are rarely part of an overall enterprise GIS.

In addition, many utilities maintain what they often call a land base for their GIS or digital mapping system. This land base consists of the land features such as street layouts, building footprints, points of interest, and other land features such as water bodies and rail lines. It is not uncommon for utilities to keep different versions of the land base for different departments. For example, a map-based fleet management system might have a digital file for the land base that comes with a fleet-tracking system. The transmission group maintains a different land base for the transmission line and the distribution group keeps a separate one as the foundation for their digital mapping system or GIS.

Often the problem is that none of the land bases match, yet conceptually, they are modeling the same information. They might have different projections, come from different sources, have different degrees of accuracy, or have no formal accuracy standards whatsoever. As noted previously, layers of information form the basis for a GIS. If the land base data set is not accurate, then the location of the electrical device layers will be equally inaccurate. In these cases, utilities find that when they want to combine different GIS data sets together, they do not coordinate. In the case of a distribution line that shares a pole with a transmission line, the two systems often will show the distribution pole in a different location than the pole in the transmission system, even though they share the same pole. If the utility then were to accurately locate the pole using a GPS device, they may find that neither the pole in the distribution GIS nor the pole in the transmission system has a GPS-accurate location.

As commercial GIS map content providers create more and more accurate digital data, utilities discover that they find it difficult or very expensive to leverage theses sources because their land base, as well as their facility locations, is incorrect.

As noted, utilities call their land information layer a land base, while today most commercial providers use the term *base map*. The solution to this base map issue is difficult, since utilities must abandon their old land bases and adopt accurate base maps or conform their land base information to known control points. In either case, they must adjust the locations of all their assets to the new or corrected land base. Conflation is the process of adjusting facilities to new locations. The end result of conflation is that the facilities are in the correct location within a defined level of accuracy. Conflation projects are labor intensive and expensive.

There are several issues with the current practice of having each department maintain their own land base, accurate or not:

- Utilities do not control the information in the land base data, so they won't know when the city or county cadastral agency changes the information unless utility personnel proactively check with the agencies. Of course, this is expensive and hit-or-miss.
- The authoritative information managed by the real estate, survey groups, and rights and permit groups have no easy way to coordinate the data with the utility's multiple versions of their land base.
- The real estate, survey, and permit group data needs to be accurate, so even if the utility wanted to incorporate the data into their land base, it probably wouldn't match.
- Utilities have to spend money digitizing changes to the land information. They are, in effect, duplicating the work that someone else (a city, town, or cadastral agency) has done already. If the utility maintains multiple land bases, the situation is simply worse.

The Simple Solution

The utility can decide to abandon the process of maintaining land base data sets and rely solely on information provided by others who either have direct responsibility for the data (such as a city or county), rely on an agency that is charged with maintaining good land information (such as a cadastral agency), or rely on a commercial base map provider that has done all the adjustments and coordination with the various jurisdictions. The real estate group, survey group, or rights and permit groups then could have ownership for the acquisition and republishing of this land base data via a web service to anyone in the organization needing core digital land information. In other words, these support groups would be responsible for the GIS for land information, which would include information they get from outside the company plus information they capture about their specific interests, like company-owned parcels or company-granted rights and easements.

This land base or base map then serves the entire enterprise and none of the operating groups has to manage it. In this way, as soon as the responsible party (the city or cadastral group) updates any aspect of the base map, the utility would automatically have access to the latest information.

Preliminary Information

Sometimes utilities need information about land features before the features are official. For example, a developer may have a preliminary subdivision plan approved and submits this plan to the utility as part of a request for the utility to design the electrical network for the subdivision. Since the subdivision is still in the planning stage and the developer has yet to build anything, the utility would need to include that data in the base map in order to do the proper planning for the electrical supply of the subdivision. In this case, the utility could create a "proposed" layer in the land base of the GIS. Since the agency in charge of the base map would provide access to the utility via an automated service (such as a Web service), the utility would have immediate access to the updated information about the subdivision. Once the new subdivision has been included in the official base map, the utility would simply delete the proposed layer. This proposed layer could show much more detail of the street construction, which would be helpful during the building of the electrical infrastructure. However, after contractors complete the street, this information is not particularly useful going forward. So the extra details needed during design and construction only need be available during that limited time. The permanent base information would consist of only information needed for ongoing operations.

Using the official base map instead of the utility maintaining its own base map solves another problem. Often a utility will only have access to the preliminary subdivision street layout and have no knowledge of any changes between the preliminary subdivision layout and the final approved as-built layout that the developer submits well after they complete construction of the subdivision. The utility could be maintaining incorrect information and not know it. It is common for changes to occur between the time a developer submits a preliminary plan for approval and the completion of construction. These could include relocation of the street alignment, shifting property lines, and the addition of easements. By incorporating official plans the utility is assured access to actual street layouts, building footprints, property lines, and easements. This will help the utility in the future when performing maintenance or working on new construction.

Utilities can separate the maintenance of the land information from the operations groups and centralize it in a support group that already has responsibility for land information. The group can serve as a broker between the outside agencies (or agencies) that have direct responsibility and control over the data. They can then provide the data to the operating groups in a way that gives them easy and consistent access to just the information they need and no more.

Different Levels of Accuracy and Display

The traditional way that utilities deal with land base is to treat the land features like the electrical and structural features. So street boundaries, building footprints, rivers and railroads are often intelligent features within the GIS databases. Yet for the vast majority of workflows, an operating group needs to see only the electrical and structural features displayed along with the land features. They do not need to edit the land base data. Utilities can structure the land base data for fast display. Caching this data in effect takes a "picture" of the core intelligent data. Displaying the results of a cached data set is orders of magnitude faster for panning and zooming than having to re-create images of streets from raw vector representations. Caching also involves establishing varying levels of detail depending on the zoom level a user needs. If a user is zoomed into a display at, for example, 1 in = 200 ft, the cached data set could show street level detail, building footprints, street labels every 5 in at a type size of, say, 8 points. If the user zooms out to a display scale of, say, 1 in = 1000 ft, then the level of detail should be less, since the user cannot possibly see a high level of detail from being zoomed that far out. The GIS displays street names at much farther apart intervals. The GIS no longer displays building footprints but more generalized footprints at this zoom level. Zooming way out would eliminate the display of many of the side streets entirely. The idea is that at each zoom level, a different "picture" of the base map is displayed. As the user zooms in, the resolution increases, but the extent of the image decreases, meaning that in effect, the same number of pixels are being displayed. With enough zoom levels, zooming in and out appears to be seamless. Behind the scenes, the GIS manages exactly what data is being accessed to be displayed.

Optimizing the Use of Land Information

A better way to manage land information is to do the following:

- Centralize the management and publishing of base maps in a support group.
- Utilize as much commercially available or centrally available (from a cadastral agency or from a commercial base map provider) base map data as possible.
- Optimize the data for display performance for all users of the GIS.
- Supplement the data with near-real-time imagery to aid in restoration efforts or emergency management. This frees up the operations group to only have to deal with information they control.
- Work arrangements with other utilities (e.g., gas, water, telecommunications, cable TV, and public transit) that serve the same area to receive their data as a web service. Optimize those layers for performance.
- Add as much data from as many sources as possible, such as municipal boundaries and overlays—wetlands, zoning, school districts, state and counties, and location of sensitive areas.
- Add generic services like geocoding, street-level routing, and real-time weather and traffic services so that if the users needed these layers, they could easily turn them on.

Users can easily manage this data in a private cloud, where each user can access the information from any device wherever they are. In effect, the company could repackage a number of data sources and services and publish these services as web services that the users from any department could utilize. One of the key departments that would need this data is the environmental group.

Environmental Issues

Utilities involved in all phases of the business have to be concerned with their impact on the environment. Certainly, utilities that generate power from fossil fuels are intimately involved in emission management. Most will have sophisticated air monitoring equipment. Governments require utilities to rigorously report emissions and any deviation from environmental regulation. Wind farm operators and developers are concerned about bird migration, noise pollution, and strobing effects. Transmission operators are concerned about where to build lines and substations, and containing insulation oil from contaminating waterways and wetlands. They worry about endangered species nesting in towers. Distribution operations likewise need to be concerned about spills from oil-filled equipment. When siting any equipment, utilities must deal with the impact of their plans on the surroundings.

There are several areas in which GIS can be deployed to help manage environmental issues:

- *Monitoring*—since most monitoring of air, water, and noise pollution involves the location of the sources and the extent of the impact, GIS is a natural tool for reporting the current state of emissions, spills, and pollution.
- *Environmental risk assessment*—GIS can alert utilities about the risk of a nearby environmentally sensitive area by performing a spatial analysis.
- *Project management of past incidents*—GIS can also be used to manage the recovery of past environmental incidents by showing what's happening, the extent of cleanup, and the progress toward completion.
- *Site selection*—GIS can alert planners and developers of utility projects of areas of environmental concern during the planning process. GIS is a critical tool for environmental impact report, which models the project in relation to environmental impacts.

GIS and Environmental Incident Management

Nearly every aspect of the monitoring of environmental issues involves location. This management involves understanding exactly where the environmental concerns exist. For example, the utility monitors the emissions from a power generation source. The measures then determine exactly what the emissions are at that location. Utilities can map the real-time measures and see the impact on those areas. If emissions exceed tolerable levels due to a malfunction of equipment, utilities can use GIS to identify what areas are impacted and notify authorities in that area as to the extent of the issue and under extreme situations to evacuate people from the area or at least notify them of the situation. So GIS links the extent of the event to those impacted.

When a piece of oil-filled equipment ruptures, utilities must do everything they can to contain the oil. However, that isn't always possible. So they must model the dispersion of the oil and remediate any damage. Most governments will require the utility to report the event within a very small time window. The GIS is often the best tool to illustrate where the spilled oil went. GIS can pinpoint the location of a particularly sensitive area and determine if the path of the spill will reach that area. Figure 8.4 illustrates a GIS map that shows where spills have occurred.

So GIS can function as an environmental situational awareness tool that brings together monitoring of air, water, and noise contamination in relationship to the areas that may be impacted. From that awareness, utilities can make rapid decisions to notify and to respond.

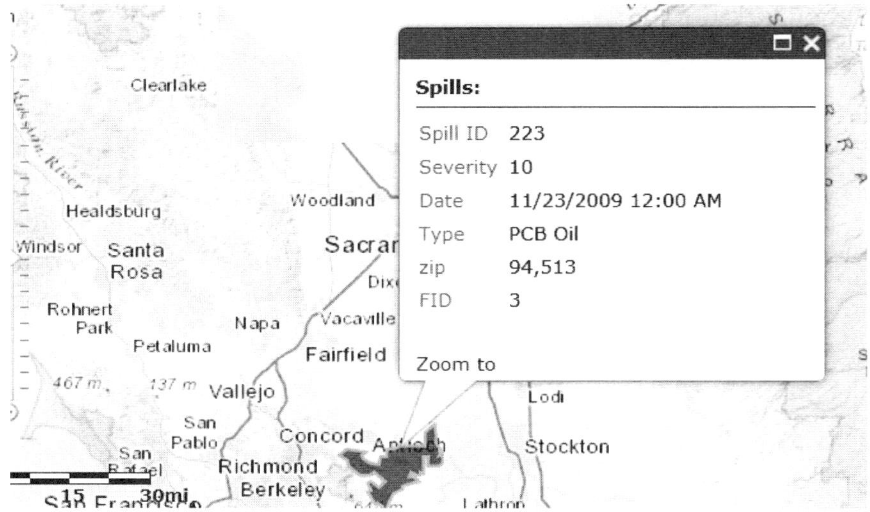

Figure 8.4 GIS map showing oil spill locations. (Source: Ersi et al.)

Risk Assessment

This book has illustrated the natural use of GIS for determining risk, which is vulnerability plus consequence. Spatial analysis and particularly weighed overlay analysis allows utilities to combine risk factors and impact factors together from a spatial perspective to determine the relative risk over a wide area. Environmental risk assessment works the same way. Spatial analysis looks at the proximity of a risk source, for example, a large oil-filled transformer to a known sensitive area, such as a vegetated wetland. The vulnerability of the source is the age of the equipment, maintenance history, evidence of leaking, history of failure, and the duty cycle. Other vulnerability issues might be proximity to salt contamination, corrosion due to high acid rain levels, and critical equipment near earthquake areas. A GIS spatial analysis model then plots a heat map or hot spot analysis of areas with highest vulnerability of an oil rupture against the highest impact of an incident (proximity to a sensitive area). The outcome of this analysis would be a risk mitigation plan that might include building higher drainage berms around the equipment or installing retention basins. The analysis helps utilities focus their environmental risk mitigation activities to those areas with the highest vulnerability and the highest negative consequences—in other words, mapping the environmental risk profile.

GIS for Environmental Remediation and Compliance

Utilities face clean up or remediation when an environmental event occurs. Sometimes the cleanup can be fast with no long-term environmental damage. Other times, the damage can be more problematic and may require long-term clean up and continued monitoring and reporting. Utilities will then need the GIS to map the extent of the damage, show the actions they are performing to restore the area, and provide progress to the environmental regulators to assure them that the utility is complying with whatever remediation plan they and the regulators have agreed to. Often noncompliance can result in stiff fines and bad press, so the more transparent the process is, the better for the utility.

GIS for Planning in Relationship to the Environment

Planning for low environmental risk is essentially the same exercise as risk mapping for electric equipment failure or the risk of environmental damage, except the goal is to plan to minimize future risk. While every significant project involves some environmental impact, utilities can avoid costly project delays if they plan for the least environmental impact upfront in the initial planning process. If the utility wishes to build a new transmission line, as noted in Chapter 3, they could perform a spatial analysis that minimizes disruption of sensitive habitat, wetlands, or beautiful vistas. By building the environmental impacts into the site selection process, utilities can avoid some of the permitting issues.

Continuously monitoring environmental issues during site selection, engineering design, and construction can help avoid costly surprises, help facilitate transparency, and ultimately assure a better overall project.

Vegetation Management

Transmission and distribution maintenance groups manage the vegetation along transmission rights of ways and city streets. See Chapters 3 and 4 for a discussion on the subject of vegetation management, specifically in relation to transmission and distribution maintenance.

The intent of vegetation management is to control the growth of vegetation, typically trees, to provide adequate clearance between the trees and overhead electrical conductors. Utilities are looking more holistically at the subject of vegetation management. In the past, utilities have simply mowed down right of ways, clearing anything and everything down to the bare ground. Often, that had the effect of stimulating the vegetation growth. Today, utilities understand that low-growing bushes can inhibit the growth of fast-growing trees, so utilities might even plant low-growing bushes along rights of way to inhibit the growth of fast-growing trees. Rather than simply clearing a transmission right of way of all vegetation, they instead encourage growth of certain species while chemically treating others to inhibit their growth.

GIS provides a way of organizing the work of vegetation along a corridor. It allows utilities to analyze areas (polygons in the GIS) of similar growth and treatment patterns. Utilities can look at factors, such as the average rainfall within the polygon, the soil type, drainage patterns, and other environmental factors that help the vegetation manager find the right treatment, cut, and planning plan suited specifically for the area for the purpose of preventing trees from growing into or near the lines.

Vegetation management for rights of way using GIS involves a change in process from simply mowing a right of way every 5 or 6 years on a schedule to cleverly determining which areas need the most attention and have the greatest impact. Using the old cyclical approach, parts of the right of way may not need treatment that soon, while other sections of the right-of-way may need attention much sooner. Since vegetaion growth is not uniform over the entire right-of-way, using GIS to fine tune vegetation management saves money and avoids costly outages.

Vegetation management for distribution involves trimming trees along the city streets. GIS utilities find that they can better manage tree trimming contractors by segmenting the work by work area, performing audits of the work, and documenting within the GIS exactly what work was done, as well as the characteristics of the area to be trimmed (e.g., the dominant species and any unusual situations, like trees that are unusually at risk of falling into the lines). Further, the GIS can assist in better analyzing the need for certain areas to be trimmed on a more frequent or less frequent basis by performing a spatial analysis of the tree trimming work areas.

Instead of simply scheduling a 15-mile feeder to be trimmed every five years, the GIS can create areas of greater or lesser need for trimming based on factors such as rainfall, soil types, tree species, types of overhead line construction (open wire versus spacer cable), and many other factors. The same principles that the risk models have can be applied to tree trimming. Trim the areas that have the greatest risk of tree growth or encroachment into the lines and delay the trimming areas where there is less risk of tree encroachment, based on a consistent GIS spatial analysis model.

Utilities can add impact to the tree trimming plan. For example, they can assess what areas of the distribution system have the biggest impact on customer reliability should a storm come through. Or they can correlate what feeder sections have the poorest reliability or power quality performance. The GIS can identify those areas. By performing a vulnerability and impact assessment, utilities can focus their efforts on what areas to trim and when that will give them the highest payback for their vegetation management dollars. Vegetation management for some utilities can be the highest expense after labor and materials. Some utilities have used GIS for vegetation management for years, while others continue to cling to the past practice of simply scheduling vegetation management by feeder or municipality.

Logistics and Supply Chain

Movement, storage, procurement, testing, and repair of materials is the job of the supply chain or logistics organization. Sometimes the management of buildings grounds is also one of the functions of the supply chain organization. These activities are essential to the successful operations of the utility and often rely on spatial information. The three major components of supply chain are as follows:

- Materials management;
- Fleet management;
- Facilities management.

Each of these groups deals with heavily with location. Historically, GIS has not played a strong role in these organizations. The purpose of this section is to examine what role GIS could potentially play.

Materials Management

Utilities are a very material-intensive industry. Maintaining a century-old infrastructure can be a significant challenge and requires a constant supply of parts for repair and replacement. The infrastructure is growing as well. Utilities need a stream of new equipment to meet the demands of a growing society hungry for energy. Keeping track of all that material can be daunting. Not doing a good job can be extremely costly. Groups within the materials management organization include purchasing, inventory, warehousing, asset tracking, and often materials testing.

A common GIS application in materials management is determining the best location for a service center or warehouse. Utilities, particularly distribution utilities, break up their service area into districts. They build one or more service centers within each district. They store and dispatch crews and vehicles from these locations. Utilities use service centers as local warehouses, where they store a selection of material, such as transformers, wire, cable reels, switches, and poles. There, they will keep lobby stock (e.g., connectors, small hardware). Sometimes, they will locate vehicle repair facilities within the service center. These facilities are often self-contained operations centers. The issue for utilities is whether these service centers are in the optimal location. Are there too many, just enough, or not enough? Should the company consolidate or add new service centers? The decision impacts the logistics of how the company responds to routine and emergency work.

Over time, the demographics of the area served by the service centers change. What may have been an ideal location for a service center 20 years ago may not be valid today. Logistics organizations routinely use GIS to answer these questions.

Solving the Serviceability and Accessibility Problems with GIS

The GIS problem is to find the driving distance from the existing or proposed service center along all possible routes. The idea is to determine if the driving distances are too long to meet service levels. This requires a network analysis of all possible routes from the service center on out. If, for example, the utility wishes to get to all locations within a given time frame, it can create a map that illustrates the driving distances from that location. The GIS solves the problem: find all routes that are within, say, 20 minutes of each service center. The results may show overlapping areas or large gaps in the ability to serve other areas.

The GIS can determine how easy it is to get to various equipment or customers in the service territory in terms of travel time or distance or a combination of both. In effect, the GIS creates a buffer around the service center based not on a polygon buffer, but a distance buffer using the streets and the street characteristics, like speed limit or traffic lights.

Utilities then can optimize the location of service centers based on criteria such as customer population, density of equipment, and even criticality of the areas served. The GIS can provide location allocation, which answers the query: given the set of existing service centers or proposed locations, provide the best response times to cover the community, as well as find the optimal cost for servicing an area.

Fleet Management

Utilities have large fleets. Many of their vehicles are heavy duty, expensive to maintain, and require care in handling and dispatching. Most heavy duty utility trucks have poor fuel efficiency. Crafting a fleet management strategy to lower miles driven while still performing the same amount of work is essential. Work can range from delivering materials to visiting customers to repairing or reading meters (the old-fashioned way). GIS provides the basis to solve the vehicle routing problem—not just finding the shortest path from one location to another, but minimizing the total miles driven for a fleet or a work group. Figure 8.5 illustrates a GIS with the optimal routing identified. The benefits of fleet optimal routing are many. The fewer miles the fleet drives, the fewer accidents will occur. By reducing the total travel time, the wear and tear on the vehicles decreases. Finally, shorter total distances traveled translate into lower labor costs. People will accomplish more work during the day. Travel time can represent a significant portion of a field worker's day.

Navigation and Automated Vehicle Location (AVL)

Utilities simply need to know where their crews are. Crews also need to know where they are in relationship to where they need to go. AVL systems use GPS to identify the location of each vehicle in real time. Since the GPS locations are also available within the GIS, a dispatcher can receive the vehicle's GPS location and the GIS can display the location of the vehicle. Each time the GIS refreshes the display, the GIS redisplays the location of the vehicle in its new location. So the dispatcher sees "in effect" the motion of the tracked vehicle in the GIS. While AVL systems are effective, they can be more effective if they are integrated with the enterprise GIS. Most often, the AVL system (even if based on a GIS-like system) only shows standard base map information such as streets and standard points of interest. Integrating AVL technology with GIS allows dispatchers and crews to see utility assets

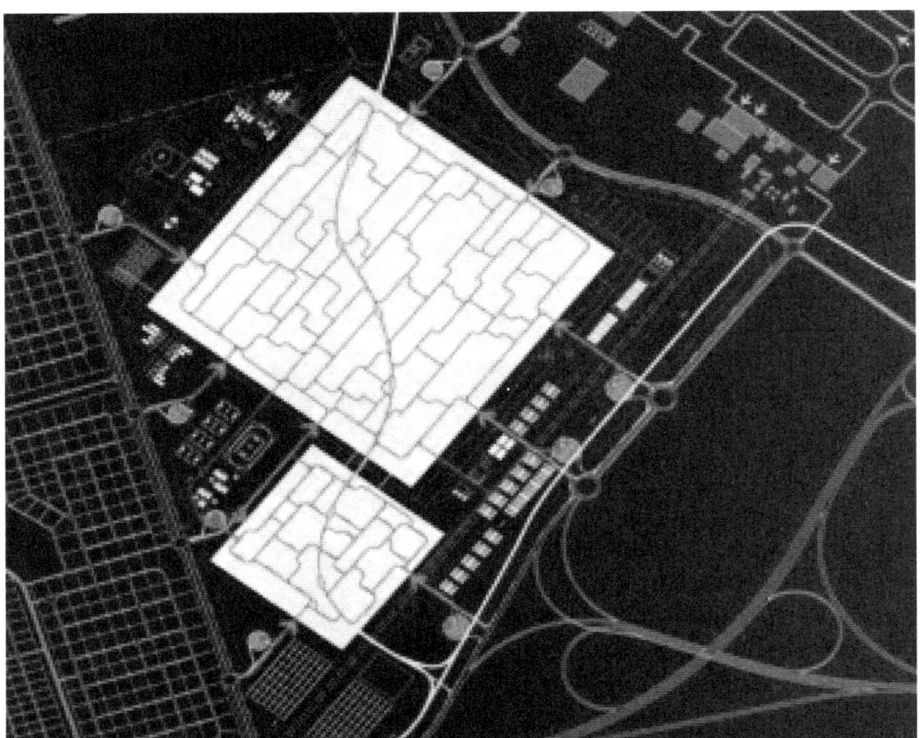

Figure 8.5 GIS used to illustrate the best route.

along with the vehicle location. In effect, the utility assets become points of interest. So a crew could navigate from their current location to a damaged pole, where the pole is discoverable within the AVL.

Since vehicles are critical utility assets, the integration of AVL technology with the enterprise GIS adds these critical assets to all the other assets the utility must manage within one geospatial technology.

Vehicle Routing Problem

The GIS solves the problem of how to sequence and schedule a fleet of vehicle visits to a group of locations taking into account certain time windows that the locations are available to be visited. For example, assume that a fleet of meter repair crews needs to visit customers to repair, test, or replace their meters. The utility has prearranged a time when the customer or electrician is at the site. The idea is to complete the routes with the available crews within the time imposed by the schedule. This assumes specific driver shifts, speed limits, and any obstacle the dispatcher knows about. GIS solves this problem by the simultaneous processing of the routing in an iterative manner, converging on the optimal solution of routes and visits. Other factors can be included, like the efficiencies of the specific vehicle.

The GIS can determine the best route taken by a utility crew. It could be the shortest, the quickest, or the one that gets the vehicle closest to the depot or service center. The GIS effectively assigns an impedance (not electrical impedance) to the route. Time or distance can be an impedance, depending on what parameter the dispatcher wants to minimize.

Closest Facility

Another common routing problem is the closest facility solver within the GIS. Here, the problem is to find the closest crew to a trouble location, like a blown fuse. Once the closest facility (a trouble shooter crew, for example) is found, the next problem is to find the shortest path (distance or time) from the problem location to the nearest crew. Figure 8.6 shows the results of a closest facility solver model.

Routing in real-life situations needs to take into account real-time situational data, such as the status of traffic lights, traffic itself, rainfall, road repairs, and bridges out of service. The analysis can also take into account the time of day the routing is taking place (e.g., at night when traffic is light or during the summer when visibility is clearer).

Asset Tracking—Fixed and Mobile

GIS can help utilities keep track of where fixed and mobile assets are located. For example, a utility may want to track the location of hazardous material they are shipping from a warehouse to a disposal facility, noting exactly when there is increased environmental vulnerability along the route.

During major outages, when winds, downed trees, or floods damage many pieces of equipment, utilities can lose track of what material went where. Crews can be more concerned about getting the power restored than worrying about exactly which transformer or switch they installed. However, later, during the recovery

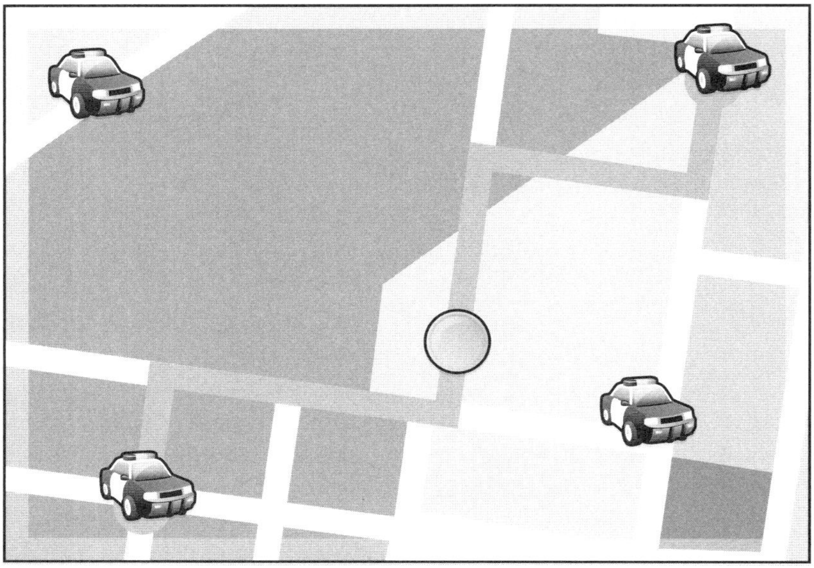

Figure 8.6 GIS map shows the results of a closest facility solver.

stage, utilities need to know the disposition of each piece of critical equipment. Using GIS to track assets can help. Radio frequency identification (RFID) tags can be included as an attribute of the asset.

Utilities use the concept of geofencing for asset tracing. Dispatchers create a polygon within the GIS that they use to demarcate an area they want to track. They may want to make sure material stays within the area, and they want the system to notify them when material enters or leaves the area. This technology is often combined with GPS tracking and RFID technology. Figure 8.7 illustrates the notion of geofencing.

Facilities Management

Utilities own and operate a number of facilities in addition to the assets dedicated to power generation, transmission, and distribution. They have service centers, warehouses, storage yards, office buildings, vehicle repair facilities, and storage sheds. They own some and lease others. How they manage facilities could well determine how well the entire operation works. Utilities use GIS to help them manage these valuable assets.

As noted throughout this book, many utilities view GIS as an electrical network documentation system. That is, its use is limited to dealing with electrical facilities, such as cables, wires, poles, and transformers. Utilities will use the GIS to help manage those assets, identifying where the assets are and what their condition is and determining what they need to do about them. However, utilities also need to make sure that they properly maintain support facilities, and they likewise need to know whether to repair or replace components of these facilities. GIS is a tool that can help utilities organize their maintenance and asset management of support facilities. For example, they need to keep track of the facilities maintenance activities of the facilities maintenance crews and contractors. They need to know where

Figure 8.7 GIS map showing the concept of geofencing.

work is going on and what impact that work has on their operations. They need to clear brush, plow snow, landscape, remove graffiti, paint and replace windows and doors, repair roofs, and a host of other activities. The GIS can provide a way to visualize and organize that work.

Facilities maintenance is part of supply chain and logistics. As such, it is critical to be able to view how support facilities play a role in the electric operations. Electric operations needs to have visibility into what support services have been scheduled, should conflicts occur.

Green and Space Planning

Utilities can use GIS to help utilities in their space planning. This can be for both personnel as well as equipment. They may need to discover where best to stage equipment in advance of a weather event or a major project. Utilities have employees that need office space. GIS can help with employee space planning as utilities attempt to optimize their use of office space and assure that employees are as close to their work as possible.

While the vast majority of emissions from utility operations come from carbon sources of generation, utilities must also be aware of their use of energy for support services, particularly in their use of energy for their facilities, such as office buildings, warehouses, and service centers. GIS can help utilities track and report on their progress toward green buildings and energy management within their facilities. They can use GIS to analyze their energy use and potentially discover ways to save energy within their own support facilities. Since utilities have many employees in office spaces, they need to manage the cost of these spaces by making those spaces as energy efficient as possible.

Finance, Accounting, and Corporate Management

While utilities rarely use GIS outside of operations, they may find that seeing financial, accounting, and corporate management information in a spatial context can be helpful. For example, a utility may want to know where the company is spending money on operations or maintenance so they can better determine the reasons for spending in certain areas compared to others.

The term location analytics describes a process whereby business intelligence information, such as spending on maintenance or new construction, can be displayed in the form of a digital map. The map can show areas of spending based on asset investment, per customer, per district, or even by time intervals. Once the GIS captures or creates the information, then it can publish that information as a service, like a web service or a map service. Then the GIS can merge or mash up other data to help the utility uncover the underlying factors that might have contributed to the variations in spending across the region served by the utility. The analysis is quite similar to other spatial analysis processes for routing a transmission line or detecting where a distribution utility is at risk during a storm.

So financial managers can gain insight into not only the where of the spending but perhaps even the reasons for the variation in spending. A simple analysis might show the variation in spending on maintenance correlated with weather patterns or wind conditions. Salt spray along the coast will have a much more corrosive impact on equipment near the coast then further inland. While business intelligence systems provide consolidated reporting of metrics, adding a spatial context often provides additional insight into what is really going on.

Plant Accounting and Taxes

A fundamental requirement of a regulated utility is to accurately report the value of its assets. There are several compelling reasons for this:

- Transparency to investors;
- Rate setting;
- Property tax accounting.

Utilities need to account for their assets for transparency to investors and for the accuracy of their balance sheet. In the U.S., for example, the Security and Exchange Commission has strict reporting rules to protect investors. Thus, utilities are obliged to accurately report on their investments. Not having an accurate accounting of their plant can cause embarrassing audits and fines.

State or federal authorities approve rates proposed by the utilities. For example, every state in the U.S. has a public utility commission that does exactly that for the investor-owned utilities that operate with the state. An important factor in rate setting is the determination of the rate of return granted the utility on its investments based on the value of its assets. If a utility files a rate case, they must be able to prove that they have correctly accounted for all their assets and the valuation of their assets. Knowing when they installed a pole or wire is critical. If they underreport their asset base, they face getting a smaller rate relief than they are entitled to.

Overestimating their asset base will result in overcharging their customers, which if discovered could lead to serious consequences.

In most places in the world, utilities pay property taxes to local, state, and federal governments based on the assets they have installed within a particular taxing jurisdiction. If utilities over report their assets, then they will pay too much in property taxes, which simply ends up as an unnecessary expense on their income statement. If they underreport and are audited or discovered, they become responsible for paying back taxes.

The ideal situation is that utilities record their plant assets consistently and with a single source of data. The GIS is the ideal location for the collection and documentation of the utility assets. As noted in Chapter 4, utilities use GIS for designing new facilities within the overall work order process, so the GIS keeps track of what assets the utility has added and retired. They can extend this process to automate the task of keeping an accurate count of assets. One of the most fundamental functions of the GIS is to count things, within areas. Linking the GIS into the design process as outlined in Chapter 5 creates a documented process as utilities add new and retire old assets. This makes the documentation clear and the validity of the equipment counts defendable.

Some utilities find that the records as recorded in their plant accounting systems do not match those held in their engineering systems (e.g., on their operating maps). By using the work order process and the GIS, the GIS becomes the authoritative source of asset data. Once the utility posts within the GIS that they have installed and put into service an asset, the GIS can publish a service or app that lists which asset has been added or retired, including data about when it was installed originally, where it is located, and the date and time they utility installed or removed the asset. The plant accounting system can then accurately determine the value of the plant at any point in time.

Sales Tax

In some jurisdictions, utilities have to collect sales tax on the use of electricity. The sales tax will depend on a number of factors, many of them spatial. For example, a school district, a water district, a special political boundary, or any number of special districts might determine sales tax. Some of these districts might even overlap. A hard way to determine the sales tax is to have the billing system carry complex tables of where each customer location is compared to a particular tax district. The GIS handles districts as a matter of routine. The billing system can issue request to the GIS passing an address of a new customer and the GIS determines which tax districts the address is located within. The billing system can then simply apply the rate associated with the set of districts the GIS returned to calculate the sales tax on the use of electricity.

Revenue Protection

The term *revenue protection* refers to theft of current. This can be a serious issue in many parts of the world, with some nations seeing illegal connections siphoning off more than half of its potential revenue. In most developed nations, theft of current is a much smaller issue but still can be a significant loss of revenue. There are two parts of this concern. The first is gaining an understanding of the problem itself. The second is in the enforcement and prosecution of the people stealing the electricity. For most utilities, particularly those with very high theft, they do not have a good handle on the extent of the problem or where the problems are most severe. Here, GIS can help frame the problem, identify where theft is greatest, and perhaps use the GIS as a tool to uncover some of the underlying problems. For example, if there are areas of extreme poverty, the GIS can identify those areas and correlate those areas with a high degree of theft. Enforcement in these cases may not be appropriate. Rather, documentation to relief agencies or appeals to the government may be the best approach. The GIS helps utilities make the compelling cases.

For the more routine cases of theft of current, the GIS in combination with smart meter data can easily pinpoint where consumption at the service level doesn't match consumption at the supply level. The GIS can also use other sources of data such as presumed vacant property that shouldn't have consumption but the data says otherwise. Theft of current drains the resources of an electric utility. Having a rational plan of recovery using the tools of GIS can help utilities gain control of the problem and focus their efforts on those areas that offer the greatest chance of success.

Safety

The safety of employees and the community is of great concern to all electric companies. There are at least two ways that GIS can play a role in the company's safety program. The first and most obvious way is to visualize where accidents and safety issues have occurred. Utilities can form the company's safety performance into an online story map, which shows each accident, safety statistics by region or division, and even some additional insight into why safety issues may be higher in one region over another. Certainly weather might be a factor in motor vehicle accidents. The frequency of motor vehicle accidents may be impacted by street lighting (or lack of street lighting), curves in the road, treed areas versus open areas, and any number of factors that vary by location. Utilities can spatially enable their safety reports, giving senior management better insight into the nature of accidents (motor vehicle or otherwise). By combining accident statistics with network maintenance activities, utilities can discover patterns of where accidents are more likely to occur. Using temporal data in conjunction with spatial data can provide additional insight. Knowing where and when accidents tend to occur provides utilities with additional tools to gain better insight into the times and places to be most diligent.

The other important aspect of the safe operation of an electrical network is having accurate information of where all the assets are located and giving field workers ready access to the most up-to-date information. Accidents tend to occur when a combination of bad things happen at the same time. For example, utilities train employees to assume all energized equipment is alive, even though they have been told the equipment is de-energized. They are supposed to test to make sure a current-carrying device is dead before working on the equipment. In addition, they are supposed to wear protective equipment even when working on de-energized equipment. However, employees don't always test, don't always follow procedure, and don't always wear protective gear. A GIS can act as a safety net to give employees the most accurate and up-to-date information about where equipment is located. If employees are using old marked-up prints to locate equipment, they run the risk of going to the wrong location, setting themselves up for a potential accident.

Security

Most utilities have a corporate security department. The mission is to protect the network and all its assets from harm. They must monitor the comings and goings of all employees, contractors, and guests. GIS often plays a role in conjunction with security systems, which use video monitoring systems. For large facilities such as a power plant the locations of the cameras are linked to the location on a map of the facilities. Since most utilities have many facilities located throughout the service territory, it becomes difficult to monitor exactly where every monitored device is located. The GIS serves as a dashboard to guide the security personnel to the location of any security violation.

Other Corporate Functions

The term *spatially enable* has been used throughout this book. In its simplest form, the term refers to taking company data that has any kind of spatial representation, such as a coordinate, a longitude and latitude, an address, a zip code, or a block identification, and simply map the information within the GIS. For example, utilities know the addresses of all their employees. Should an emergency occur and they have to call in employees to help out, it could be helpful to know where the employees live in relationship to where they are needed or where there are hazards for their employees to drive near. Utilities can spatially enable their human resource systems. Once spatially enabled in GIS, they can combine that spatial information with other maps, such as maps of work, problems, or security breaches.

On a regular basis, senior management reviews the key performance indicators (KPIs) of the company. In the vast majority of cases, they can spatially enable those KPIs and often gain additional insight about what is going on. Often KPIs are averages. Mapping the KPIs gives management better insight into where hot spots (areas where the KIPs are showing the biggest problems) are located and what the demographics of those hot spots represent.

Utilities will do customer satisfaction surveys. While their rankings may be acceptable, it would be instructive to map where they are having satisfaction problems. Then they can combine this customer satisfaction data with other data, such as high bills or any other metric to gain a better understanding of what is really going on. Then they can craft more focused solutions to improve customer satisfaction. They may discover that their tree trimming program, while good for reliability, is creating ill will in certain neighborhoods. This is very easy to visualize in a GIS but virtually impossible to ascertain from reams of charts and graphs. Figure 8.8 shows how a utility can simply map the results of a customer satisfaction survey within the GIS. While this information is certainly interesting in and of itself, the utility could overlay demographic information to uncover what types of people tend to be unsatisfied. They could also overlay reliability data or any other data that might give the utility greater insight into how customers are feeling about

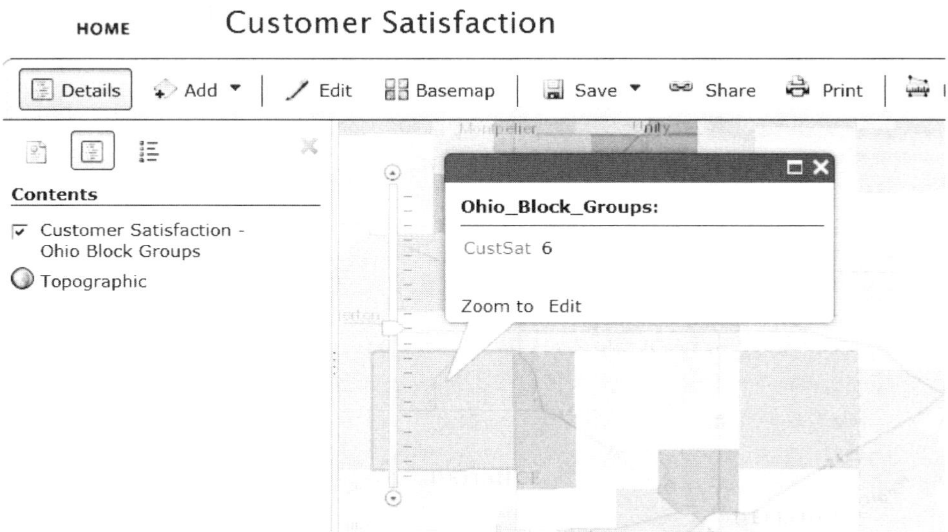

Figure 8.8 GIS map showing the customer satisfaction results. (Source: Esri et al.)

their electric service. Utilities could also overlay social media data so they gain even further insight into why customers are feeling the way they are.

GIS Is Critical to Shared Support Services

Most utilities think of a GIS as an engineering system designed to support operations. And it does. However, the same technology that determines the risk profile of a transmission line or the optimal path of a new transmission line can be used to model many other processes within an electric utility, whether or not the government has unbundled the generation, transmission, distribution, and retail businesses. As in most other companies, utilities do marketing and outreach, deal with materials and buildings, and have employees and accidents. For many of these functions, if not the vast majority of them, location matters and mapping business functions can be transformative. It allows utilities to see patterns that are difficult or impossible to see any other way.

A modern GIS is not a departmental system where data sharing is difficult. The simple task of mapping a business function throughout the entire enterprise with the ability to combine maps with different data creates a discovery process. If visualization alone is not enough, GIS can perform sophisticated spatial analysis that combines layers of information into a single map that often surprises users who see patterns that they have never seen before.

APPENDIX A

GIS 101 for Electric Utilities

Utility users need to understand what makes up a GIS, even if a cloud vendor or someone else mangers the GIS. The key to understanding the technology of the GIS is a concept called the GIS information model. While there are as many definitions of information model as there are opinions of what it means, for the purposes of this book, it consists of three components:

1. *The data model*—exactly what data the GIS directly manages;
2. *Information products*—this is the definition of exactly what the users will see;
3. *Workflows*—exactly what the GIS does to the data to produce what the user will see.

Data Model

Whether the utility manages the GIS in house completely or using SaaS (software as a service), someone has to define what data to be included in the GIS. This is the data model. As noted all throughout this book, each business unit has a unique set of things that is important for that business in relation to some kind of location. For example, a customer care organization needs to know where every customer meter is located. A transmission organization needs an inventory of every tower and its location as well as every other piece of equipment.

At the heart of the data model is the notion of a feature. A feature is anything that the GIS represents, like a street, a cable, a switch, or a lightning arrestor. It can also represent a boundary, such as that of a city or country or it can represent the result of an analysis.

There are three kinds of data commonly used by the GIS:

1. Vector;
2. Raster;
3. Network.

In each case, the GIS stores characteristics or attributes about each of the elements in these three data types.

Vector Data

If one were to view a line on a computer screen, the line actually represents a series of pixels that are displayed close enough together to give the appearance of a line. To store all these pixels would take considerable storage, particularly if the line is long and the resolution of the screen is high (meaning many pixels on the display). An alternate way to store the line would be to store the fact that the feature is a line along with parameters that define the line, such as the coordinates of the end points. Thus, there are two ways to store data about a line: in raster form or a large set of pixels or in vector form, which is a small set of values that define the mathematical parameters for the line.

The advantage of vector data is storage is small. The disadvantage is that in order to actually display the feature (e.g., the line), the computer system must convert the vector representation to raster. So that process takes computing cycles. Raster data types have large storage requirements, but the data is in the correct form to be displayed without much translation.

Vector data tends to be in the form of point, line, and area or polygon features. A fourth kind of feature is a volume feature, in which the feature can be represented in three dimensions. In all cases, the feature data is a collection of coordinate pairs (or triplets, if considering three dimensions); a point represented by one pair; a line by two; and polygon feature by as many coordinator pairs as there are vertices. An area feature by definition has to close in on itself. Area or polygon features often represent phenomena like areas of similar elevation or sensitive areas like vegetated wetland areas or corridors which represent the best area to route a transmission line.

For simplicity and ease of visualization, GIS modelers represent electrical features as one of the core data types for standard two-dimensional representation of mapping information. A pole is certainly not a point on a two-dimensional plane. However, most modelers use a single point to represent a pole feature. The point is actually located at the center of the pole on the ground. Conductors are cylinders, but most modelers abstract them as lines on the map with no width.

Vector Layers and Features

To some extent, the data model is similar to the database schema of a standard relational database management system (DBMS). A standard DBMS stores unique entities in tables. The tables are very similar to spreadsheet work sheets. Each table has a number of rows and columns. So, if a DBMS were managing utility poles, it would create something called a pole table. The table would consist of rows and columns, with each row representing a specific pole, while each column in the table would represent attributes of the specific pole. In other words, the column headings would represent various characteristic possibilities for a pole, such as pole height, class, date manufactured, and date installed. The column headings represent the possible attributes of the entity or feature. In a GIS vector data model, the same structure exists. There is a table for each feature classification.

The difference between a standard DBMS table and a GIS database table is the inclusion of geographic representation, the shape (point, line, polygon, volume), and the values for the location (sets of coordinates). Just like a normal relational

DBMS, GIS tables can have cardinality relationships (one to one, one to many, many to many). GIS tables can have relationships to other non-GIS tables as well.

While it is possible to have a table for every single variation of a feature, such as one table for wood poles and one for steel poles, most modelers group features of similar type and shape (point, line, polygon, and volume) into single tables.

The GIS data model consists of a set of tables, defining attributes, and relationships. As in most DBMSs, the GIS data model defines behavior, such as the domain or range values of certain attributes. For example, a rule would be that a 5-ft or 500-ft pole is not an acceptable value for pole height. In fact, only specific poles values would be valid, such as 35 ft, 40 ft, 50 ft, and 75 ft. The data model would not allow other values. Disconnecting devices are also modeled as point features (although, some utilities model these types of devices as line features, since they have two or more terminals), but have a completely different set of attributes. So the GIS data model will include a disconnecting device table. The disconnecting device table would include a variety of different types of disconnecting devices, such as fuses, circuit breakers, and reclosers. The model would include other tables such as are high-voltage conductors, transmission towers, guy wires, and medium-voltage transformers. The features represented by a single feature table must have the same shape (point, line, or polygon or volume).

Layers represent the way the GIS displays features. A layer displays a single feature. In a typical GIS, the software presents the user with a check box next to the layer label. When the user checks the check box, the GIS software displays the layer. A display could show a base map, the medium-voltage overhead feeder sections, and only the disconnecting devices, but not the low-voltage or transformer layers. Any combination of layers is possible for display. Most GIS software systems are smart enough to provide additional filtering to display features with certain characteristics, like voltage level or condition.

The idea is that the data model defines the complete list of features that the modeler wants to include within the GIS, plus all the attributes that needs to be included in the GIS as well as relationships of features to each other and the relationships of features to other data sources outside the GIS. An example of a feature in which data is accessed outside the GIS is a SCADA point. The GIS defines the shape—in this case a point, the location of each specific point, a relationship to other features (e.g., a substation breaker and some kind of identification attribute, say, the SCADA point identification). The SCADA point identification is the same point identification that exists within the SCADA system. Thus, the SCADA system would export a table or publish a Web service listing the real-time value or values of the SCADA point on a periodic basis. The GIS then accesses the web service or the table to retrieve the current value of the data. Thus, the GIS could display the real-time data from SCADA on the map while not storing any of the real-time data. Any event data can be treated the same way.

Raster Data

Early utility GISs did not use raster data for anything other than the scanned images of their old paper maps and records. This was not so much to take advantage of the unique abilities of raster data, but more in an attempt to avoid having to convert old paper drawings to vector format.

Raster data, as noted earlier, consists of a matrix of cells or pixels. Most people are familiar with raster displays. Television displays are series of cells, with each cell firing a point with a color and intensity. The resolution simply is the number of pixels or cells. The idea is that each cell contains one and only one point. The value of the point determines the color and the intensity or shade. When viewed from afar, the picture looks like a continuous display or image.

A typical raster data set used for a utility might be vegetation type over a transmission right of way. Each pixel could have a value of 1 to 10, where 1 represents the area with very slow growing vegetation and 10 represents an area with very fast growing vegetation. When the GIS displays this image, it has a rule that states a value of 1 is very light green, whereas a value of 10 is very dark red, with various color combinations in between. The result is a map that shows continuous color patterns from light green to dark red. The darker red areas represent the areas the utility has to perform intensive vegetation management compared to the lighter green areas, which show the areas that the utility can put off vegetation management for longer periods of time.

The data model for this data set is quite simple: the data model is the description of what the pixels represent (e.g., elevation, vegetation, or risk). It also includes a definition of what values the attributes of the pixel mean. Examples are dark is bad, light is good, red is bad, and green is good.

In some situations, raster data sets can represent discrete features. For example a river can be represented as a series of pixels in which the pixels of the river are colored while the other pixels are not.

Raster Layers

A full-featured GIS includes both raster and vector images together. Typically a raster data set is a single layer, turned on or off. For example, a customer care employee might want to look at meter consumption in relationship to the population demographics of age. The meter consumption data might be point vector information, with each point representing a single meter or service, while the demographic data is a raster data set, with each pixel representing an average age bracket of people living in the area represented by the pixel. Together, they could show a correlation of age with consumption patterns. The GIS could also convert the vector consumption data into a raster image as well, in which the pixel shows the average consumption values of all meter points that lie within the area represented by the pixel.

Network Data

A network data set is a special case of vector data and is especially valuable for electric utilities. In the case of a simple vector feature, each feature represents a shape: point or line. The fact that a point and a line occupy the same space or are very close may or may not be significant. A network data set defines that a set of features participates in a network. That is, the network data set defines connectivity. So, a disconnecting device, which may be a point feature, represents a node in a connected network, whereas a conductor represents a line in a connected network. If a data model defines both the disconnecting device table and the conductor table as belonging to the network data set, then the GIS assumes that if they share one end point, they are connected. This is the notion of GIS network data.

The application of a network data set for electric companies is obvious—the network data defines the relationship of current-carrying features to one another for the purpose of network tracing and to define the network connectivity for network calculations like load flow or short-circuit analysis. A less obvious use of network data is in structural networks, such as underground duct bank networks. These networks contain manholes, vaults, hand holes, boxes, pipes, and tunnels. These networks can be fairly complex. Planners can automate the routing of power or communication cables through the network automatically if the models define structural elements as part of a structural network.

The term raceway is a generic term for any piece of equipment designed to hold or contain cables, power or otherwise. Examples of raceways include above-ground cable trays (in power plants and substations), tubes, conducts and junction boxes (underground or mounted on walls). So even within a power plant or substation (or in fact any industrial facility), modelers can create extensive raceway networks. These networks can connect to other structural networks outside the substations or power plants. This gives utility operators the ability to determine routing throughout all of the various utility facilities. For example, a utility many want to understand the location of a medium-voltage feeder from the last manhole outside the substation, within the substation basement conduit system, to the cable tray feeding the specific circuit breaker. This would be particularly valuable to know during a mishap such as a fire at the substation. The raceway network is, in effect, the containers for the electric network equipment. The raceway network can also provide an undesirable pathway. For example, during a flood, water will flow through the raceway network. Or, the raceway network could provide a pathway for smoke, fire, or pests.

The other common use of a network data set is for routing of vehicles to project sites, warehouses, customers, and inspection locations. As noted in Chapter 8, GIS can provide significant benefit within the utility's supply chain operation. Much of the benefit comes from the use of a network data set within the GIS.

The Data Model Development

For vector data sets, there are a number of techniques for developing an electric utility data model. Most GIS vendors provide templates to get started. The process to create vector data sets often starts with a description of very high-level kinds of data (e.g., conductors, transformers, disconnecting devices, and customer locations). Then, each high-level type is broken into more specific types, such as overhead conductors and underground cables.

The raster data models often consist of the different types of raster data sets required for utility use, such as base maps of streets, topological data sets, and imagery. In each case, these raster data sets have multiple representations for various display scales. For example if the utility wanted to view the location of their facilities at a small scale (zoomed way out), the image is low resolution. As the user zooms in, the GIS uses a different raster image at a much finer resolution showing more detail, but at a much smaller extent. As the user zooms further in, the GIS calls an even finer image, showing just a few blocks of data, but at a very sharp resolution. Eventually, the GIS reaches the highest resolution data set. As the user zooms in further, the image blurs, since the pixels simply get further apart. Technically, there is no limit to the detail with which the GIS can display imagery. The only limit is the availability of an extraordinarily high- resolution imagery data set. Of course, the greater the resolution the larger the data set is (since the data for each pixel has to be stored) and the more expensive and difficult it is to capture.

Modelers create network data sets by assigning certain features to a named network, such as medium-voltage network or a secondary network duct bank network or a pipe type cable transmission network.

Information Products

As stated earlier, an information product is what the user will see. In the not-so-recent past, the user would see only output products from their GIS. These were a set of standard map products that utilities created with the GIS or in the past on paper. In the vast majority of cases, utilities printed the maps on a plotter. In the old days, utilities printed the maps on translucent paper so they could be copied using blueprint technology. Once copied, utilities shipped the maps to the various departments. Of course, this process was awkward, time consuming, and expensive. The most obvious problem was that as soon as a department received their new paper print, the print was out of date.

While paper maps are still common, utilities are migrating to the use of mobile technology (laptops, tablets, and smart phones). Then, the user gets the most up-to-date version of the GIS. However, this change may not be particularly transformational, since the output product might well be the same image as the paper map, just more current. The more subtle and transformational change is the idea not so much of output products, but of information products. That is, the users get just the right information they need at that time for the job they need to perform.

Designers of the original paper maps (a century ago) created all-purpose maps. They had to, since they had to draw each map by hand. They couldn't afford to create hundreds of specialty maps for every conceivable process or workflow the consumers of their maps might need. So they invented standard map products, like one-line diagrams, feeder maps at standard scales or grid maps. Each of these maps contained as much information as possible so as to make the map-making process as efficient as possible. With the advent of digital mapping, the new mappers replicated the old maps, only digitally. They then printed them to look exactly like the old maps.

With the advent of mobile devices, users continued to demand the same look of the old maps on their mobile devices. The good news or bad news, looking at it from different perspectives, was that mobile devices were limited in the amount of information they could reasonably show. The good news was that designers of mobile information products, while limited by display capability and screen real estate, were not constrained by the variability or variety in what they could show. For example, an old-fashioned feeder map displayed labels for a number of equipment parameters, whereas a mobile map user can simply hover over a symbol for a transformer for example and see its kVA, voltage, or impedance rating. This created the opportunity to create an infinite number of displays.

This led to the notion that information products were not defined by a strict set of layout requirements but were more about what information *could* be presented to users, not what information *is* presented. Thus, as displays become less cluttered, the information presented becomes significantly enriched and more sharply focused. *The shift is from standardized static maps to interactive maps where the information created is on user demand.*

The transformation from standard output products to information products means that what people see is simpler, more relevant to their job at hand, and more effective and more consumer like.

What are common information products for utilities? They really are driven from use cases for common tasks that utility workers perform. For example, suppose workers have a job to make sure that the utility complies with a state or federal regulation that all pad-mounted transformers must be inspected every other year. They can use a simple GIS tablet app that tracks the location of their vehicle, creates a buffer around their location, searches for a pad-mounted transformer location (even if it is buried in snow or hidden from view), and queries the work management system (integrated with GIS) to see if the pad-mounted transformer was recently inspected. If the answer is yes, then nothing happens. If it is no, the app alerts the users that they passed a pad-mounted transformer that they need to inspect. The app displays the transformer and a simple form filled in with only the attributes required for the inspection, along with the location of the transformer. The workers then stop their vehicle, discover the equipment, do the inspection, complete the form, and are on their way. The users could also take a picture or video of the equipment as well. Behind the scenes the app collects the inspection information, wirelessly transmits the new information to corporate, and delivers it to the work management system. In this case the information product is simply a map that displays the workers' location and a transformer only if the transformer meets a certain requirement—in this case, one that they need to inspect. This same information product scenario could be applied to any inspection-related activity.

The key to the information product is the use case—what specific tasks do workers need to perform? For example, the distribution planners need to analyze feeders for future loads. Repair workers need to inspect and repair equipment. In the old days, the distribution planners and the repair workers might have used the same cluttered output products. Since the GIS is an information system, it can serve information to meet the specific needs of the user.

The information product is linked closely to the use case:

- Who is doing what?
- What is the exact information they need to perform their task and no more?
- For each step along the way, what visualization do they need that helps them to proceed to the next step?

The key is what information they need to do their job. Too much information will slow the workers or the process down. Too many buttons or options creates confusion and increases training.

Workflows

The third component of the information model is the set of workflows that the GIS applications need to perform. There are thousands of tasks in an electric utility that could involve the use of spatial information. Many of the major ones are documented in the main body of this book.

A workflow links the data model to the information product.

In the previous example of the inspection process, there were several workflows—most of them automated. The GIS facilities information model of course includes a complete inventory of every electrical piece of equipment, structural equipment, and customer equipment, all linked to their correct spatial position. The GIS data model also includes specific identification tags for each piece of equipment to which the GIS could link other systems within the utility, such as the work management system. The workflow for this simple inspection app is as follows:

- Compare the inspector's vehicle location to the known location of all pad-mounted transformers.
- Query the information contained within the GIS facilities model for pad-mounted transformers only at a location within a fixed buffer distance.
- Once the GIS app discovers a pad-mounted transformer within the buffer zone, query the work management system (all behind the scenes using some form of integration technology).
- If it discovers that the equipment complies with the inspection criteria, the app does nothing (or performs some other function, like notifying the inspector of some special situation—for example, there was a fault on the secondary of the transformer and it might be worth checking anyway).
- If the transformer is not in compliance, the GIS app alerts the inspector. The app highlights the exact location of the transformer on the map.
- Presents the inspector with a panel indicating what fields the inspector must fill in to complete the inspection. It might even show the prior inspection results.
- The inspector completes the inspection, fills in the form, and indicates "Complete." If a photo or video is required, the inspector can take a picture or video. The picture or video is linked with the inspection report.
- GIS app automatically records the date and time of the inspection.
- Transmits the data to the corporate network where the appropriate systems are updated with the current information.

Of course, there are variations on this workflow. For example, if the inspectors discover a pad-mounted transformer that is not in the GIS or one that is in the wrong location, that would trigger a new app with a new workflow. This is a very simplified example of an inspection use case.

For each of the business applications, utilities need to create use cases that define each step, what information is needed for each step, what visualization is needed at each step, what the user needs to do (for example fill in a form or take a picture), what the app will do after the user does something and what visualization needs to happen then. This process is repeated for each step in the use case.

Spatial analysis

One of the most powerful workflows a GIS performs is spatial analysis. There are perhaps hundreds of spatial analytic algorithms. Some of them are fairly complex. However, most companies can easily implement them if they have the data. The most common for utilities consists of the following:

- *Queries*—where is something?
- *Counts*—how many are there?
- *Relationships*—where is something in relationship to something else?
- *Buffering*—what is surrounding something?
- *Overlays*—how can we combine disparate spatial data sets?
- *Union and intersection analysis*—how do things relate or not?
- *View shed*—what impact does terrain and elevation have on viewing?

Queries

A spatial query can be as simple as finding the location of something by an address, ID, or equipment number. An example is to have the GIS find the location of a pole by its pole ID or pole number. The user types in the pole number and clicks. The GIS app zooms to the location, highlights the pole, and perhaps highlights some other aspect of the pole. Maybe it has a street light attached or was recently replaced. More complex queries are indirect. For example, "Find all the lightning arrestors within District A that were made by Acme Lightning arrestor and installed between April 1996 and March of 2003." This query looks at all the lightning arrestors in District A, then narrows the search to only those that meet the query requirement.

The result of a spatial query is a selection set (i.e., the set of features that meet the requirement). Often in a spatial analysis, the GIS saves the selection set and uses it for further analysis. Suppose, for example, that the reason for the previous query was that Acme Lighting Arrestor Company's products manufactured during a certain period failed more regularly during high periods of lightning or in areas more susceptible to lightning. The selection set can be combined with a lightning frequency and intensity map to alert the utility to replace those arrestors only in the severe lightning areas. The idea is that after the query, a selection set of arrestors is available for further analysis.

Counts

Utilities will often want to know how many of a certain type of equipment are located within certain boundaries. For example, a utility may want to know how many street lights they operate within a municipal boundary or how many street lights have not had their lamps replaced over the last three years within a certain maintenance area. The result of the analysis is not a map, but a number. Plant accounting departments would like to know how many poles were damaged during the last storm so they can account for the replacement costs. GIS performs spatial analysis to determine how many things are included where.

Relationships

Utilities want to know how far their equipment is from something else. Often, that "something else" is something to be avoided. For example, many utilities operate vehicle repair facilities. Some utilities even have body and paint shops. They need to know how close these facilities are to sensitive environmental areas, such as vegetated wetlands. Should a spill occur, the utility needs to know well in advance how fast to respond and how best to mitigate the risk of damage and resultant fines and bad press. There are still concerns about the impact of electromagnetic waves to health. Utilities can determine if there are sensitive populations or facilities in relationship to transmission lines.

A relationship analysis might be "show me the distances from all nuclear power plants to population densities."

Buffering

Buffering is similar to relationships except that a buffer analysis determines a fixed distance from a feature. Once the GIS establishes the buffer, the spatial analysis determines what is contained within the buffer. The common use of buffer analysis is the answer to this question "find all the property owners within a certain distance from the proposed location of a new substation project boundary." The buffer creates a boundary a fixed distance from the substation parcel. The GIS then produces a selection set of only those parcels contained within the buffer. An option of the query would be either completely contained or partially contained within the buffer zone. The result of the query is a list of parcel owners within the buffer that must be contacted. The resultant selection set would have to contain parcel numbers that could be linked to the parcel owners.

Buffering can also be used to determine if there are any zoning or other kinds of land overlays within a work zone. For example, some jurisdictions require a special permit to do work within a certain distance from protected lands, such as habitat protected areas or even burial grounds. Buffer analysis can help uncover areas that the utility may want to avoid, whether performing design, site selection, or looking at where to route transmission lines.

Overlays

Overlay analysis is one of the most powerful aspects of GIS and spatial analysis. Society has used the concept of overlay analysis for centuries to help people visualize the relationship of one factor to another. The idea is to create a series of scaled drawings of something, each drawing created on a transparent layer. When an analyst lays the drawings on top of each other, they can see the relationship of the various factors. For example, if crime analysts wanted to gain a better understanding of certain patterns of crime, they could create a street map and then create a transparent layer of where certain crimes have occurred over the last several years. They could then create an additional transparent layer of where known drug activity was happening and another layer of where abandoned houses were located. When all the layers are lined up and layered on top of one another, the analytics could discover if there was a correlation of certain types of crime with poverty or abandoned property or drug use.

With a fair amount of work, the analyst could create a new map that illustrated where the three factors of a certain type of crime, together with drug use and abandoned property come together. In effect, GIS does this automatically by using the process called overlay analysis. Before an overlay analysis can be performed, the data sets must be converted or have the same coordinate system.

In concept here is how it works for raster data sets:

1. Each pixel in the data set has a value. For each data set, the GIS must normalize the values to the same scale. For example, if analysts wants to create a risk assessment using a flooding zone data set and lighting intensity data set, the GIS must first convert the flood map from different flooding depths (feet) to an flood intensity map (on a scale of 1 to 10, for example). The GIS must create a lightning intensity map (again on a scale of, say, 1 to 10) from a basic insulation level (BIL) scale to a lightning intensity scale. In each case, the number 10 represents the highest possible risk, whether lightning or flood risk. The GIS does this process for each data set in the analysis.
2. Now each pixel value for each data set is based on the same scale. The GIS then simply adds up the values of each pixel in the same location, which are lined up geographically.
3. The GIS then creates an additional data set based on the combined pixel values and displays the results.
4. The analysts can optionally assign a weighing factor to each data set, so when the GIS adds up the values for each pixel, it uses the weighting factors. This allows the analyst to perform a sensitivity analysis based on a combination of different weights for each factor.
5. The GIS (or user) can assign different textures or colors to the map data set based on the value of each of the pixels. The common practice is to assign dark colors to the areas that have the highest pixel values.
6. The final data set then can be stored as a new data set and used in subsequent analysis.

If in fact, there is a confluence of factors within a certain area, the resultant data set will have different areas with different colors or shades of colors. This type of map is called a heat map, since the result (if there is a correlation of factors) will show the most intense areas in the darkest colors.

The GIS can also perform this analysis using vector data sets. For example, suppose analysts want to discover which older underground conduits lie in areas of unstable soils. The GIS stores the values of the underground conduits in a vector data set, whereas the data set for unstable soils is a raster data set. The GIS would convert the vector data set to a raster data set. The age of conduits would be stored with the pixel value that represented the location of the conduit. The GIS would then normalize the resultant conduit raster data set on a scale of, say, 1 to 10, with 10 being conduits older than say 50 years old. The GIS would then normalize the soil stability map having a similar scale. So the overlay analysis would simply add the pixel values. If an old conduit pixel shared the same location as an unstable soil pixel, the resultant pixel value would be 20. So the resultant map would show a conduit map, with at-risk conduits (old and in unstable soils) painted in dark red.

Union and Intersection Analysis

GIS can combine different data sets using a process of union and intersection. This analysis is similar to overlay analysis. If, for example, analysts want to understand where two different demographic groups overlapped in their customer opinion surveys, they could perform an intersection analysis. In this case, GIS discovers where the values within each data set are the same. Again, the data sets have to be normalized to the same values. Instead of adding the pixels, as was done in overlay analysis, the pixels are checked so see they have the same value. If they do, then the GIS creates an new output data set that includes only those areas that have the same pixel value.

Unions work in a similar way. A union data set attempts to combine two or more different data sets into one. So, depending on what the analysts want to accomplish, the union process combines pixels of one data set value or range to a data set range of the other data sets and creates a resultant data set in which areas of interest are combined together.

View Shed Analysis

A common problem for utilities is visualizing what a proposed facility will look like after it is built. This is a common question asked during public hearings. The view shed analysis requires an input raster data asset. The concept is that each cell or pixel within the raster data set is assigned a value (normally a 1 or 0) depending upon if an observer, such as a neighbor near a wind turbine, can see each point of the raster data set. The resultant data set illustrates what if anything the observer can see.

View shed analysis is also helpful in the determination of wireless coverage for smart meters that use cell technology. The smart meter is the observer, and the location of the receivers are part of the raster input image. If the cell tower or receiver cannot be seen by the smart meter, then the utility must find an alternative location or additional location for the receiver.

Learning More About Spatial Analysis

This presentation of spatial analysis is meant only to convey the ideas behind the concept. Perhaps the best source on spatial analysis is the excellent three-volume series by Andy Mitchell, *The Esri Guide to GIS Analysis*, which is referenced in the bibliography.

Coordinate Systems and Accuracy

It goes without saying that any GIS should start with some form of standard coordinate system, so that utilities can take advantage of widely available data sets. GIS can perform transformations from one coordinate system to another. Some utilities built their GIS from base maps that have no specific coordinate system. They will find it more difficult to perform spatial analysis, since the data sources they could get from the other sources will not line up with their data. Different levels of accuracy can co-exist if they are all based on the same coordinate system. This is not the case if the utilities created their base map or land base information from some arbitrary coordinate or no coordinate system.

If utilities want to integrate their workflows with GPS tracking, then they must migrate from their old legacy mapping systems to a coordinate system consistent with GPS.

Cloud and Web-Based GIS

Over the last several years, the web has allowed many users to publish their map-based data. This means that utilities can take advantage of a huge number of data sets hitherto hidden in file cabinets of governments, commissions, organizations, and agencies. These data sets range from environmental overlays to soil types to crime statistics (utilities can use these as well) to flood and sensitive areas to zoning maps. Other utilities, such as water and sewer utilities, can publish their utilities over the web to be consumed by the electric utilities. Utilities can access accurate base maps so they don't have to maintain their own. Likewise utilities can publish their work over the web for use by other agencies (with proper security, of course). The era where data has to be stored locally to be used in the GIS is over. Modern GIS can consume data sets created by others stored in the cloud easily. GIS can understand the data by reading the metadata (data about the data) so that it can easily consume and transform the data sets.

This will make spatial analysis more powerful and more common. Utilities will be able to share spatial data sets across their organizations like people currently share social media data.

More Study Is Required

Volumes and volumes of technical work have been written on the technical aspects of GIS. The reader is encouraged to dig deeper into the use of GIS for improving utility workflows. There are many other examples of spatial analysis that can be applied. Only the most common examples are cited here. The purpose of this appendix was to give a conceptual understanding of GIS, not to provide precise workflows. Certainly every GIS project has its own unique way of implemention. The purpose here is to give the reader an appreciation of the most common uses of GIS for use by the electric utility industry.

APPENDIX B

The Facility Model, GIS, and Smart Grid

Smart grid has a number of different meanings, and the term is often overused. Many utility people would argue that the electric grid is in fact very smart. Smart essentially means automated. When a problem occurs on the electric transmission system, such as when a high-voltage line comes too close in contact with a tree, the line flashes over. What happens after that happens automatically. Relays detect the rapid change in current and take virtually instantaneous action. Huge circuit breakers interrupt the high current in a number of locations, isolating the faulted transmission line. If for some reason the breaker fails to operate, a stuck breaker scheme starts and opens additional breakers to clear the fault and stop the flow of fault current. Failing to do this quickly and automatically can create a situation of instability leading to all kinds of catastrophic situations. So the transmission system is quite smart.

Smart Grid Automates the Distribution System

Distribution substations have smart logic that triggers when something bad happens as well. Breakers open, systems transfer loads, historian systems and data warehouses record and store events. These smart systems have been around for decades. Utilities have installed more and more automation at substations using modern digital technology to replace the older analog systems. So why is there all the discussion today about the smart grid if the grid is already smart? What isn't so smart in today's electrical network is what's going on in the medium-voltage and low-voltage networks. Many utilities have been totally blind to what goes on with an individual customer in real time or even within hours.

Smart Meters Are the Heart of the Smart Grid

The vast majority of smart grid work involves the conversion from simple electromechanical meters to smart meters. Utilities around the world are installing smart meters. Smart meters measure electricity consumption and other factors like voltage and power factor at very small time intervals. The intervals can be seconds. Utilities have been using interval meters for years for large commercial and industrial users. However, in these cases, the meter stores the interval readings locally in the meter itself. Simple residential meters typically are mechanical and simply record an accumulated consumption. Utilities commonly read interval meters and simple meters monthly using various techniques by sending people to the meter location to get the data.

Smart meters use telecommunication technology to transmit the meter reads automatically and are able to receive messages from the power company to the meter to take an action. That action may involve taking additional readings or simply shutting off the electrical supply. Most smart meters are able to communicate to other devices such as customer equipment directly.

Advanced metering infrastructure (AMI) collectively describes the smart meters plus the telecommunications technology and the systems to process and capture the information. A meter data management (MDM) system processes the meter data for billing and dissemination. For many people, an AMI is essentially the technology of smart grid. Smart meters can notify utilities of a power failure, momentary interruption in power, and voltage sag, and some even can detect harmful harmonics in the lines.

Aside from the obvious benefit of not having to send meter readers to the field every month, the real long-term benefit of the smart meter is to be able to control the consumption—to be able to turn on and off individual loads inside a customer's facility for the purpose of demand control. Or, in the simplest application, to simply shut off the supply of power on the day the customers move out or when they don't pay their bill.

AMI Will Introduce Dynamic Pricing

Another aspect of AMI is the future ability of utilities to provide dynamic pricing of power. Say, for example, the wind is blowing like crazy, the weather is mild, and it's on a weekend. Even during peak times, the cost of the power may actually be quite low due to increased use of wind power and solar energy. On the other hand, during the hottest part of the day, dispatchers predict there could be a shortage of power, and the pricing of power may be quite high. Without interval consumption data, the utilities and the customers have no way of knowing whether they are using or conserving during high or low consumption times. Today, the total energy used in the billing period (typically one month) determines the customers' bills. With smart meters, utilities can link the price of power with the cost of power on a very short time interval. Further, customers will be able to query their usage to better manage their cost of power.

Self-Healing

Another element of smart grid is the installation of equipment to help make the distribution network smarter and more responsive to problems. Since smart meters are sensors, they can help determine what's going on in the medium- and low-voltage networks. In addition, as part of the smart grid initiatives, utilities are installing additional sensor devices and automated switching systems in the medium-voltage network and maybe someday even in the low-voltage network to respond much more quickly to power failures.

Without human intervention, given the smart meter data, the medium-voltage sensors, and automated switches, the medium-voltage network can reconfigure itself (like the transmission system does today) to minimize the number of customer out of power. Utilities call this self-healing. Today field troubleshooters and dispatchers switch and reconfigure the network manually. Self-healing processes can restore large numbers of customers in seconds automatically.

What Really Is Smart Grid?

So smart grid really means a few different things:

- Demand response between customers and the utilities with the ability of customers and utilities to measure consumption in a very short interval and respond when necessary to control or manage consumption.
- Automation of the medium voltage system to restore customers much faster. With the addition of sensors on the network, the utility can do a better job managing the voltage of the lines to optimize power flow and lower losses.
- Automation of the integration of renewable resources and energy storage of the system.

The idea is that the smart grid is the merging of two networks, the traditional electric network with an intelligent communication system for the purpose of managing consumption and improving reliability.

What Does GIS Have to Do with Smart Grid?

The smart grid needs the facility model of the electric network to figure out what the sensors are communicating. The GIS provides the single source of information about the configuration and data of the electrical network. Since the smart grid adds new equipment to the electrical distribution system, it will become more important to understand the relationship of this new equipment to the existing grid. The facility model described in Chapter 4 serves as the basis for nearly all smart grid initiatives.

In addition to the facility model created by the GIS, the ability of the GIS to perform additional data manipulation and analysis provides greater insight into the smart grid. The GIS helps to identify where gaps might exist in the communication coverage of the smart meters to helping utilities understand consumption patterns. The GIS performs spatial analysis to determine where utilities can focus their efforts on demand response and conservation programs. Figure B.1 show a GIS map displaying smart meter locations.

GIS Issues Dealing with Smart Grid

Three issues will inhibit the full usefulness of the smart grid related to GIS. The first is the incompleteness of the electric distribution facilities models within the GIS itself. A smart grid needs to understand exactly the phase connectivity of each element and the exact representation of the low-voltage network. Many utilities have not captured the specific phase information.

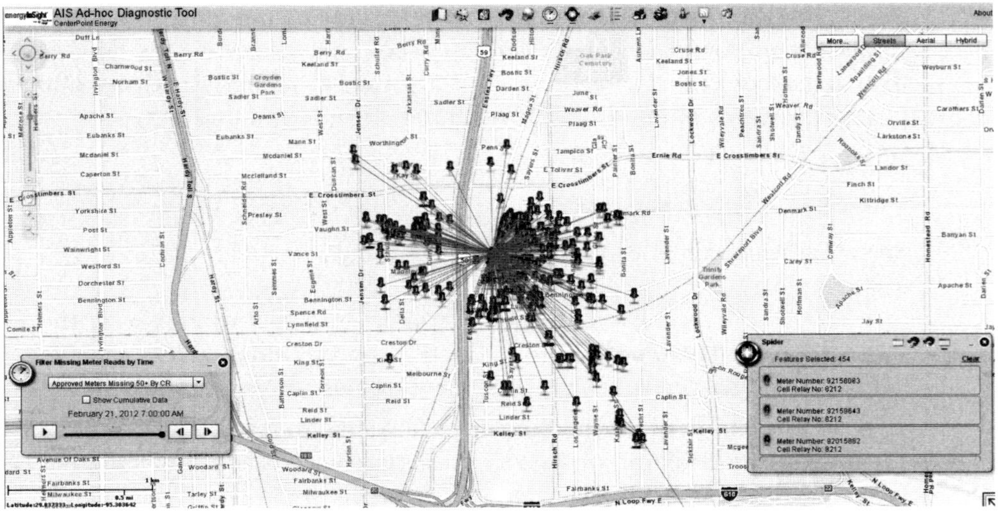

Figure B.1 GIS map displaying meter installations. (Source: Centerpoint Energy.)

The second issue is the timeliness of the data in the facility model. Even if the utilities had a complete representation of the facilities model, if the workflow to keep the data current takes too long or there is a backlog of unposted changes or unresolved conflicts to the GIS, then the smart grid will be operating on imprecise information. If the smart grid automatically opens and closes switches, it will need to know the exact representation of the network, not some representation that is a week, month, or, worse, several months old. The notion of self-healing, of automatically reconfiguring the network during a power failure, requires an up-to-the-minute representation of the network coming from the GIS, so the workflow of updating the network needs to be streamlined.

Finally, as noted earlier, some utilities built their GISs on base maps that had no standard coordinate system, so the GIS does not represent the electrical and structural facilities in their correct position. So utilities cannot use GPS in conjunction with the GIS to locate equipment. This will be more critical as utilities install more and more sensors. Location of these sensors will be critical.

Smart Grid Is also About Renewable Energy

The mix of generation is changing. Today, utilities generate the vast majority of electricity from large centrally managed power plants, fueled by nuclear power, coal, water power, and oil. In relative terms, utilities understand exactly how to manage these power plants. When the demand for power increases, the smart systems of these plants adjust their output. When the demand falls, the plants respond in a controlled automated way. In the near future, utilities will produce more and more electricity from variable sources of power, like wind and solar energy. When the demand increases, the utilities can't just add more wind fuel to wind farms or turn up the sun to provide more solar energy. While the utilities have a very smart way of dealing with the present mix of power plants, they will need to figure out a smart way of dealing with power sources that are far more unpredictable. As a result of the drive toward renewable power, utilities will install more and more energy storage facilities to help mitigate the variability of these new generation sources. Further, customers, developers, and utilities will install many of these new renewable sources of power to the medium- and low-voltage networks, which currently have the least automation, monitoring, and control.

Utilities will need to understand the exact relationship of all renewable energy sources to fully determine the impact of shifting generation and demand on their distribution systems.

A Grid Cannot Be Smart Without GIS

The notion of smart grid implies that the grid operates on information not on intuition, observation, or customer calls. To be smart, the utility must have workflows in place to quickly capture any changes to the grid, be able to differentiate a power failure from a communication failure, and have the right analysis tools in place to provide situational awareness of the various smart grid systems. GIS provides the needed data management, analysis, and awareness to help the smart grid really be smart.

Summary

In the preface, I wrote about one of my bosses, Russ, the president and COO of an electric and gas utility. Russ thought that simplicity was the key to success. In the preface, I listed the few guidelines that Russ liked to follow. He said, "if you follow these guidelines, you will run a successful utility."

- Make money for the shareholders.
- Keep customers happy.
- Keep employees safe.
- Obey all laws and regulations.
- Keep the lights on.
- Respect the environment.

His last guideline, which I liked to call the "goldenest" rule was:

- Keep us out of trouble and out of the newspapers.

I also shared my recollections with several of my coworkers at the utility. Nick strongly believes in simplicity and consistency. Recall from the preface that Nick wrote a memo that became so trusted for its wisdom and simplicity about relay protection that it earned the simple name of *Nick's Memo*. This was presumably his only memo. I contrast Nick with Dan, who loved technology, but made things so complicated that maintenance and upkeep of his complex substation schemes cost the company a fortune. Johnny, Nick's predecessor, followed Nick's philosophy to the letter, and never had to write a single memo in his long career.

What do Russ, Nick, Dan, and Johnny have to do with enhancing electric utility performance with GIS? GIS provides the simplest and most elegant form of communication: displaying complex business and engineering situations in the form of a map. Maps are as old as civilization. Maps are about discovery. From finding buried treasure to charting mineral deposits on Mars, to seeing voting patterns on cable news to locating the best place to site a coffee shop or a nuclear power plant. While the underlying technology of GIS can be very technical, the concept is simple and consistent; GIS organizes information by location, analyzes the results, and creates awareness of the results in the most common user interface known to humanity—the map.

Make Money

So Russ outlined the main stakeholders in his simple guidelines. His first one was to make money. Russ understood that if a business enterprise didn't make money, it would fail. A quick examination of any utility's income statement discovers where they make and spend the bulk of their money. They get their money from customers, either wholesale or retail customers. Improving collections, reducing theft and losses, and better marketing of conservation programs improves their revenue. Understanding where losses are, where people aren't paying bills, where are people most likely to steal electricity, and where to market gives utilities gives better insight into the effectiveness of their revenue enhancing programs.

Utilities spend the bulk of their money on fuel (either directly or purchased), people (including contractors), and materials. GIS improves productivity and material use by organizing the data and giving utilities the tools to better assess whether people are working more effectively. This is done by understanding where parts of the system are at most risk and which parts of the system need upgrading. GIS is fundamental to asset management.

The common theme of each of the chapters in this book is the concept of spatial analysis. Taking the map's ancient user interface and coupling it with modern data management and display technology, GIS discovers the hidden treasures. In this case, the hidden treasure is not marked with an X, but simply a display showing where productivity is low, where materials are wasted, or where crews are taking longer to find things. Without the map, no one finds the hidden treasure.

The notion of making and saving money applies to all parts of the business, from generation of energy supply to transmission, to distribution to emergency management to customer care to shared services. Every one of these organizations has processes that depend on location, some more than others. Ignoring location handicaps the organization.

Keep Customers Happy

For years, utilities referred to their customers as ratepayers. Not long ago, when customers wanted electric service, they had to fill out an application, like applying for college or a mortgage. Maybe the bank would grant the loan or the college would admit the applicant to college, if the applicant was lucky. Today, most utilities are very concerned about customer service. This is even truer today after unbundling, with parts of the energy supply chain deregulated. Even the regulated elements of the business need to be very concerned about happy customers. Today, consumer, business, and political groups motivate utilities more than ever. Unhappy customers can result in a violation of Russ's guideline about making money. In extreme cases, regulated utilities might even lose their franchise.

All throughout this book, I make the case that GIS discovers the best route, the places of greatest risk of failure, the places where customers are most unhappy, where to apply programs, and even how to discover the underlying causes of why customers are unhappy by performing what GIS professionals call *mash ups*. The powerful concept of layering different kinds of spatial information today and performing an overlay analysis gives utilities insight as to where things are most severe. In this case, the "X marks the spot" is not where the treasure is located, but where a confluence of customer events and impressions direct a utility to make changes.

GIS helps keep customers happy by mapping the pulse of the customer.

Obey All Laws and Regulations

Utilities, of course, make every attempt to obey all laws and regulations. However, things happen, accidents occur, and violations of laws and regulations result. In many cases, the violations are location based. A misguided line truck drives over a vegetated wetland. An unfortunate back hoe operator digs into a 345 kV pipe type cable. A substation transformer ruptures, sending insulating oil into a stream. A power plant mistake causes a plume of soot to blanket a residential neighborhood.

In most cases, each event has a spatial component. The notion of proximity is extremely helpful in avoiding many accidents. GIS routing, distance to nearest facility, environmental risk assessment, and simple counts of areas where utilities have violation issues are all standard GIS functions. Obeying laws and regulations also involves an understanding of where potential issues might arise, like using spare fibers in transmission lines for commercial use, over easements that do not permit such use. In these cases, GIS can help utilities avoid violations that may not be so obvious.

Keep the Lights On

Perhaps the most fundamental rule of success in running a utility can be the most challenging. Keeping the power flowing. The three main capabilities of GIS are data management, spatial analysis, and providing awareness, as detailed in the introduction. Keeping the lights on requires the gathering of all kinds of information about where facilities are located, what their specific condition is, and what other factors might contribute to an equipment failure. Using spatial analysis, utilities can combine any number of data layers to clearly discover where the risk of failure is greatest, and then visualize the results. During major emergencies, as detailed in Chapter 5, GIS can provide awareness of where customers are out of power, and how best to respond to the emergency by creating a heightened awareness of factors that may make restoration more difficult, such as flooding areas or landslides are downed trees or impassible roads. GIS's use of imagery can provide enhanced understanding of the situation by performing change detection between pre- and post-event images. Instead of relying solely on field investigation, utilities can arm themselves with intelligence to better respond to power failures, so when the lights do go out, they will have the tools to respond and rebuild in the best possible way.

Of course, the best case is to harden the power system so as to avoid power failures. GIS can pinpoint areas where hardening provides the biggest bang for the buck. GIS can help utilities mitigate risk, be ready, rapidly respond, and effectively recover in the event of a major power failure.

Respect the Environment

Nearly every aspect of power generation and delivery involves challenges to the environment. Power plants create emissions. Wind farms may endanger certain birds and migratory patterns. Coals storage can harm water aquifers. Gas pipelines supplying turbines can explode. Transmission lines create electromagnetic waves causing issues. Many transformers are oil-filled, which do rupture. Utilities have lots of heavy-duty trucks that generate emissions. Many have garages with hazardous materials, which if not cared for, could cause harm.

The key for utilities is to fully capture the location of the potential environmental source of harm (like the location of transformers) and the location of any sensitive environmental area. GIS connects the dots between the source of potential damage and the at-risk area. This is a classic GIS problem. Knowing beforehand the environmental risks gives the utility the tools it needs to mitigate the risk. Should an accident happen, the utility will have the information to deal with the risk in a timely and efficient manner.

Keep Us Out of Trouble and Out of the Newspapers

A variety of factors create most accidents or problems. For example, accidents are often the result of a number of failures, such as not knowing where a sensitive environmental area is in relation to a project, or not assessing the potential of a substation being flooded during a 100-year storm. The common expression that ends up in the newspapers after some troublesome situation is, "the utility should have known." Unlike any other technology, GIS adds a spatial context to data that pinpoints where problems can happen. GIS is really about the *where* of a utility. With an enterprise GIS, where information is shared throughout the company, utilities gain a better understanding of the connection of all of the factors. Utilities are notoriously run as independent silos of departments. The idea that the utility should have known is often a result of one department not knowing what the other is doing. GIS provides a common framework of understanding: the map.

GIS is not simply about creating maps that used to be created by hand. Yes it does that, but it is so much more. As stated a number of times, it is about discovery and fully understanding where the utility's biggest problems might be so that the utility can take action to deal with the problem effectively or to avoid it altogether.

GIS Technology for Enhanced Electric Utility Performance

GIS helps utilities see the operations in a simple and consistent way in the form of a map. This book breaks down each organization of the utility into its major components. It illustrates the value of location to the business. Nearly every aspect of the business involves location. GIS helps utilities solve their biggest problems in each of the areas of their company by helping providing a simple, elegant, and ancient context to their business—taking the complexities of the business with all of its factors and metrics and finding where X marks the spot—on the map.

Bibliography

Arctur, D., and M. Zeiler, *Designing Geodatabases*, Redlands, CA: Esri Press, 2004.

Burke, J. J., *Power Distribution Engineering Fundamentals and Applications*, New York: Marcel Dekker Inc., 1994

DeVries, W., "How GIS Supports Outage Management," *Electric Power and Light*, Vol. 88, No. 5, September, 2010.

Ferrey, S., *The New Rules: A Guide to Electric Market Regulation*, Tulsa, OK: PennWell Corporation, 2000

Foresman, T., *History of Geographic Information Systems: Perspectives from the Pioneers*, Upper Saddle River, NJ: Prentice Hall 1997.

Fu, P., and J. Sun, *Web GIS: Principles and Applications*, Redlands, CA: Esri Press, 2011.

Gelllings, D., *The Smart Grid: Enabling Energy Efficiency and Demand Response*, Liburn, GA: The Fairmont Press, 2009.

Glasgow, J., "Siting Linear Facilities with Geographic Information Systems," *Paper Presented at the ESRI International User Conference,* San Diego, CA, August 2004.

Huxhold, W. E., and A. G. Levinsohn, *Managing Geographic Information System Projects*, New York: Oxford University Press, Inc., 1995.

Kersting A. P., J. Kersting, C. F. Filho, and M. Müller, "Airborne LIDAR and GIS Tools in Transmission Line Re-Rating Projects," *ESRI International User Conference,* San Diego, CA, July 2005.

Lakervi, E., and E. J. Holmes, *Electricity Distribution Network Design 2nd Edition*, London: The Institution of Engineering and Technology, 2007.

Meehan, W., "As Seasoned Utility Staff Retire, Will They Take Wisdom With Them?" *PowerGrid International,* Vol. 16, Issue 7, July, 2011, pp. 28–32.

Meehan, W., "GIS Makes Smart Grid Smart," *GeoInformatics*, Vol. 12, Oct/Nov, 2009, pp. 7–9.

Meehan, W., "GIS: A Gold Mine of Opportunity," *Transmission and Distribution World,* June 2011, page 20.

Meehan, W., "Transmission Risk Assessment Using Spatial Analysis," *Proc. Transmission and Distribution Conference and Exposition (T&D), IEEE PES,* Indian Wells, CA, May, 2012, pp. 1–5.

Meehan, W., *Empowering Electric and Gas Utilities with GIS*, Redlands, CA: Esri Press, 2007.

Mitchell, A., *The Esri Guide to GIS Analysis Volume 1: Geographic Patterns & Relationships*, Redlands, CA: Esri Press 1999.

Mitchell, A., *The Esri Guide to GIS Analysis Volume 2: Spatial Measurements & Statistics*, Redlands, CA: Esri Press, 2009.

Mitchell, A., *The Esri Guide to GIS Analysis Volume 3: Modeling Suitability, Movement, and Interaction*, Redlands, CA: Esri Press, 2012.

Pansini, A. J., *Guide to Electrical Power Distribution Systems Fifth Edition*, Tulsa, OK: PennWell Publishing Company, 1996.

Public Utilities Reports, Inc. *Principles of Public Utilities Operations & Management*, Reston, VA: P.U.R. Guide, 1999.

Shoemaker, T., and J. Mack, *The Lineman's and Cablemans'* Handbook, New York: McGraw Hill, 2007.

Short, T., *Electric Power Distribution Handbook*, Boca Raton, FL: CRC Press, 2004.

Sorebo, G. N., and M.C. Echols, *Smart Grid Security: An End-to-End View of Security in the New Electrical Grid*, Boca Raton, FL: CRC Press, 2012.

Tomlinson, R., *Thinking About GIS, Geographic Information System Planning for Managers, Third Edition*, Redlands, CA: Esri Press, 2007.

Vadari, M., *Electric Systems Operations, Evolving to the Modern Grid*, Norwood, MA: Artech House, 2013.

Zeiler, M., *Modeling Our World, 2nd Edition: The ESRI Guide to Geodatabase Concepts*, Redlands, CA: Esri Press, 2010.

About the Author

Bill Meehan is a leading authority on the application of GIS to utility work flows. He has authored numerous articles, delivered many keynotes and papers around the world, and won several speaking awards. Bill is the author of three other books, *Modeling Electric Distribution with GIS, Empowering Electric and Gas Utilities with GIS,* and *Power System Analysis by Digital Computer.* He is a frequent contributor to the online magazine *Energy Dimensions*. He is well known for his storytelling style and thought-provoking messages, as well as his unique ability to make complex subjects easy to understand. Bill is busy at work on his next book, which looks at the energy future for the United States.

Currently, Bill is the director of utility solutions for Esri, the world's largest GIS technology company. He has been with Esri for more than ten years. Bill is responsible for marketing and the strategic global direction for the utility, transportation, and telecommunication business segments. Prior to joining Esri, Bill was the vice president of electric operations for NSTAR, an electric and gas utility headquartered in Boston, Massachusetts. There, he provided executive-level leadership for all aspects of electric transmission, substation, and distribution operations, including maintenance, construction, vegetation control, system operations, and GIS. In 1999, Bill comanaged the merger of the Boston Edison Company with Commonwealth Energy to form the company called NSTAR. In 1998, senior management tapped him to restructure systems and processes in anticipation of electric utility deregulation. Bill spent many years as manager of engineering at Boston Edison. It was while managing engineering that Bill conceived of and led the Boston Edison GIS project to successful completion.

Bill earned a master of engineering in power systems from Rensselaer Polytechnic Institute and a bachelor of science in electrical engineering from Northeastern University. He completed the executive management program at the Sloan School of Management at MIT. Bill taught a number of engineering courses at the Graduate School of Engineering of Northeastern University and at the School of Engineering of the University of Massachusetts. He is a registered professional engineer in the Commonwealth of Massachusetts, and a member of IEEE and Eta Kappa Nu.

Bill has been heavily involved in community work, having spent ten years on the Planning and Zoning Board in Easton, Massachusetts. He served the board of directors of Easton Cooperative Bank, was a member of the Easton Lion's Club, chaired the United Way Campaign for Boston Edison, and served on the advisory board of Future Engineers, an outreach program for inner-city high school students. Bill currently serves on the board of directors of the YMCA of the East Valley in southern California.

Index

I

K

L

M

N

O